Henry Wynmalen's

HORSE BREEDING
&
STUD MANAGEMENT

Revised and enlarged by
Ann Leighton Hardman

J. A. ALLEN
London

British Library Cataloguing in Publication Data

Wynmalen, Henry, b. 1889
    Wynmalen's horse breeding and stud
    management.
    1. Livestock : Horses. Breeding. Stud farms
    I. Title   II. Hardman, Ann Leighton
    III. Wynmalen, Julia
    636.1'082

ISBN 0-85131-453-8

First published in 1950
Reprinted 1971
Reprinted 1973
Reprinted 1975
Reprinted 1976
Reprinted 1995
Revised and enlarged edition published 1989

Published by J. A. Allen & Company Limited,
1, Lower Grosvenor Place, Buckingham Palace Road,
London SW1W 0EL

Printed and bound in Hong Kong

To
BASA

foaled in 1931 at Count Esterhazy's Stud at Tata,
Hungary, from a long line of impeccable ancestry,

a true Equine Aristocrat, with the heart of a lion,
the pride of a King and the gentleness of a lamb,

a Champion in his own right, never defeated under
saddle, and the Sire of many Champions,

a great ride, known to thousands, and the Sire of
great performers, known to thousands,

a true, generous and unforgettable friend.

Preface to the Revised Edition

Henry Wynmalen's 'Horse Breeding and Stud Management' has long been regarded as a classic in its own field, containing as it does the author's lifetime practical experience of this subject. Well used copies of the book are to be found on the library shelves of most horse and pony breeders throughout the world, for whom it has become almost a bible.

However, nearly 40 years have passed since 'Horse Breeding and Stud Management' was first published. The last 10 years or so, alone, have seen what can only be described as a remarkable revolution in the science and practice of horse breeding. In the light of these dramatic changes, it was considered essential that the book should be up-dated before re-publication.

As far as possible I have retained, undisturbed, the chapters describing the author's own ideas on breeding. Elsewhere out-dated theories have been removed and detailed descriptions of modern methods introduced. These include: encouraging early conception; hygiene on studs; pregnancy diagnosis, including use of the ultrasound scanner; disease conditions and their prevention, especially equine metritis and virus abortion; modern thoughts on foaling and care of the newborn foal; up-to-date theories and methods of feeding brood mares and young stock.

Now available in its new and modernised form, this book will continue to be 'the bible' for stud personnel of all ages.

ANN LEIGHTON HARDMAN

Contents

Foreword

TO A KEEN horseman there is no greater satisfaction than can be found on the back of a horse — with the sole exception of the added thrill that comes from a horse that has been bred at home.

The most brilliant performers that I have bought have never thrilled me half as much as those that I have bred. Turning hounds for the first time on a home-bred youngster, sailing with him over the first big fence, are memories that can never die.

That is why, for years, I have bred most of my own horses. That is how, for years, I have learned, and loved, the fascination of breeding, its difficulties, and its delights, the always entrancing charm of young life.

I have found, in my home-bred horses, an amount of charm, of intelligence, of gentle spirit, of confidence, and of undoubted friendship, that one does not often find in the less fortunate animal that has gone from hand to hand.

That has pleased me all the more, since the experience gained could only confirm me in the methods of training horses, based on that understanding and confidence between man and beast that I have always advocated.

I believe that the methods used on my stud farm may well be of interest to other breeders. These methods are, on the whole, in conformity with those in use by leading studs of Thoroughbred horses, and they are specifically applicable to that type of breeding and stud management.

But I have compiled the information in this book with the special view in mind that it should be useful to the breeders of all types of horses, whether their breeding is carried out on a large or on a small scale. To me Thoroughbreds, Arabs, Hunters, Polo Ponies, Children's Ponies, Hackneys, Farm-horses, are all of interest; they are horses all, and all are entitled to receive the best care which our knowledge and circumstances permit.

Whereas some of the views expressed, and some of the information provided, may be of sufficient interest to give experienced breeders food for thought, the book had, of necessity, to provide first and foremost for the not so experienced breeder, anxious to study the subject. There are today a great many private breeders who breed or rear on quite a small scale, and the enthusiasm of many of those is not always matched by their knowledge. If my efforts can be helpful to them I shall feel particularly gratified.

I am indebted to Lady Wentworth for the outstanding photographs, taken by herself, of Arab Horses reproduced in this book.

My thanks are also due to all other owners concerned, and to the various breed societies, who have helped me so kindly with the loan of pictures of their best animals.

It will be clear to readers that this is a treatise on practical stud management, and not a work with veterinary or obstetric pretensions. For that reason a bibliography is given of leading veterinary works to which readers may refer if they wish.

1950   HENRY WYNMALEN

List of Photographs

Heavyweight hunter *Darrington*.

Modern heavyweight hunter *Flashman*.   (Bob Langrish)

Middleweight hunter *Beau Geste*.

Modern middleweight hunter *Dual Gold*.   (Bob Langrish)

Champion show hack *Liberty Light*.

Modern Champion show hack *Rye Tangle*. (Bob Langrish)

Modern show cob *Buzby*.   (Bob Langrish)

Champion Hackney 1938 *Barcroft Belle*.

Modern Cleveland Bay *Masterful Jack*.   (Carol Gilson)

Polo pony *Valentine*.

Modern riding pony youngstock Champion *Bradmore Nutkin*.
(Carol Gilson)

Modern working hunter pony *Towy Valley Moussec*.
(Carol Gilson)

Modern Welsh mountain pony stallion *Coed Coch Rhion*.
(Carol Gilson)

Modern New Forest pony stallion *Deeracres Franco*.
(Carol Gilson)

Modern Exmoor pony stallion *Dunkery Buzzard*.
(Carol Gilson)

Modern Dartmoor pony stallion *Allandale Vampire*.
(Carol Gilson)

Modern Shetland pony Champion *Bincome Venture*.
(Kit Houghton)

Modern Highland stallion *Dallas of Stanford*. (Carol Gilson)

Modern Dales mare *Brymor Mimi*.   (Kit Houghton)

Modern Fell mare *Bewcastle Bonny*.   (Carol Gilson)

A foal's first outing.

First outing: the effect of the warm sun on its back.

It will suddenly jump.

First outing: the dam will watch her child.

The softness of the turf.

First outing: galloping about in circles.

Back to mother.

The first leading lesson.

Leading the foal with a stable rubber.

Foals should be friendly and confident.

Taking more than normal interest in a nervous foal.

Arab mare *Shabryeh* with colt foal by *Riffal.*

The farrier's visit.

Weaning time.

Yearlings at grass.

Yearlings in winter.

Photograph Credits

The Publishers would like to thank the photographers Carol Gilson, Kit Houghton and Bob Langrish, for their help in supplying modern photographs to complement those reproduced from the original edition of *Horse Breeding and Stud Management*. Acknowledgements to each photographer appear in parenthesis in the list of photographs (see p. xv).

List of Text Illustrations

CHAPTER ONE

The First Essential

THE first essential for anyone intending to breed horses, on
however small a scale, is a stud farm. Possibly only a stud
farm of sorts, consisting maybe of a few fields and some
buildings and not necessarily the elaborate kind of establishment that is usually associated in the public mind with the
term stud farm, as the home of some Derby winner, or of
similarly valuable bloodstock. But small or large, we must
have facilities in the way of land and buildings, if we wish to
breed horses and rear them. And, small or large, these
facilities must be as good as it is possible to make them, if we
mean our breeding and rearing to be a success. If, as is more
than likely, the facilities at our disposal are not in every
respect as good as they should be, there are ways and means
of improving things. It is part of the interest and of the
pleasure derived from breeding to improve one's stud in
every way, and to maintain it in the best possible conditions
of health, fertility and cleanliness. When anything is worth
doing it is worth doing well.

In the first chapters of this treatise, therefore, I will deal
with the problems of farming one's fields and of arranging or
adapting one's buildings for the purposes of a stud. The
principles set out apply to any size of stud, whether of five or
of five hundred acres, with the proviso, of course, that one
must cut one's garment in accordance with one's cloth,
which implies that the person with but little land must needs
forgo some of the possibilities and advantages enjoyed by his
fellow with a larger area at his disposal. But the principles
themselves do not vary, and it remains true, of course, that
the man who proposes to breed on a small scale only can
achieve remarkably satisfactory results on a small acreage.
Nevertheless, breeding is by its very nature an expanding

1

kind of undertaking, since foals become yearlings and they in turn grow into two-, three-, four- and sometimes five-year-olds before they are disposed of, and if meanwhile new foals keep on arriving it is obvious that we are faced continuously with the problem of accommodating an ever-increasing population, which may at times become quite alarming. It is as well, therefore, to have a little elbow-room for expansion, and excepting for the owner who wants to breed only one or two foals, and no more, which can be done on quite a small place, it is advisable to have rather more land at one's disposal. Small-scale breeding can be done quite efficiently on forty or fifty acres of good land, but a bigger area is always more economical; anything from one hundred to two hundred acres makes a very nice place. Whilst that is the sort of acreage that I have had in mind in compiling the material that follows, I would repeat that much of what is recommended can be applied on a much smaller scale, and all of it, quite obviously, on a larger one.

The Land

IT IS generally conceded that horses do best on a limestone subsoil, since such land contains the required elements of calcium and lime whereof the horse stands in need for the formation of bone and tissue.

But horses can be reared satisfactorily on almost any type of land, provided it be maintained in good heart, fertile, clean, healthy and well drained. Good drainage is really the first consideration, since satisfactory conditions of fertility and cleanliness cannot be achieved on waterlogged land. On such land, which is acid, or sour, horses cannot thrive.

Certain types of sandy, chalky or gravelly soils are self-draining and present no problem. But large parts of the British Isles consist of heavy, or very heavy, clay. Clay is a plastic, almost solid substance consisting of very fine particles which stick together very tightly; it contains little or no air; and although it can absorb a certain amount of water it becomes completely saturated in wet seasons. In that condition it is impenetrable to further rainfall, with the result that the topsoil becomes waterlogged, excluding all possibility of a healthy plant life.

In such cases adequate drainage must be restored, or provided, before any substantial improvements in fertility can be achieved. This entails good, clean, substantial ditches and, in most cases, surface drainage by means of piped drains or, in suitable conditions, by means of mole drains. The

3

provision of really effective drainage is nearly always a matter of considerable expense; but it is an expense fundamental of good husbandry, and one that will always pay a dividend and that will prove an economy in the end.

Apart from lack of drainage, many of our soils suffer from deficiency of certain mineral components that are essential for a healthy plant and animal life; deficiency of lime, phosphate and potash are fairly general. It is a simple enough matter to determine such deficiencies by a chemical analysis of samples of soil taken from different parts of the farm.

Most fertilizer manufacturers will take the soil samples from your fields and have them analysed free of charge. They will then recommend the most suitable fertilizer from their range. This should take into account the land use as well as any phosphate, potash or lime deficiency. For instance, a stallion stud will need an early-bite to accommodate the visiting mares, whereas a private stud may need grass in the mid-season when their mares and foals come home.

The advisory service of the Ministry of Agriculture, Fisheries and Food, A.D.A.S., will give you an unbiased analysis and report, but they do make a charge of £5.75 per sample.

Apart from the valuable and precise data that can be established only by a soil analysis, much useful information may be gained from observing the natural plant growth in a given meadow. If this consists of an abundance of fine leafy grass with an admixture of clovers, and with a comparative absence of weeds, the inference is justified that not much can be wrong with such a meadow; that is particularly true if growth throughout the field is fairly even, and the herbage of a deep green colour. Absence of clover denotes lack of phosphate; matted grasses, moss, mare's-tail, sorrel, buttercups, denote acidity or lime deficiency.

Grass is, on any stud- or stock-farm, by far the most important crop; it is the mainstay of any type of stock-farm; the animals pass the major part of their life on it, derive from it the bulk of their feeding, and depend for their development in the first place on the quality of the grass whereon they feed.

4

If for that reason alone, we want first-class grazing; but there are other important reasons, since the best grassland will carry twice the number of stock, and will feed them better than indifferent grassland could; the best grass will grow earlier in the season and last until later in the autumn; it will recover much quicker, when given a rest, so that it can be grazed more often and for many more days or weeks during a season.

Now the grassland at our disposal may be good, moderate or bad. Bad grass cannot be got into first-class condition by any form of management, however excellent. In such cases the best, quickest and most economical method is to plough, to make a clean seed-bed, and to reseed with a mixture of first-class strains of grasses and clovers suited to local conditions.

Moderate grassland can often be improved beyond recognition by good management and by good manurial treatment.

Management of grassland implies the method of preparing the land for grazing, or for haycutting, of grazing at the right time, for the correct length of time, and with the appropriate number and the right kind of animals, of resting it, and of the treatment to give it during periods of rest.

Meadows required for early grazing in the spring, or for haymaking, should be shut up and kept free from stock from the previous autumn.

If the land be in need of lime, this should be applied around November, so as to give the autumn rains ample time to wash it in; grazing animals take a considerable amount of lime out of the soil, and unless such soil happens to be naturally rich in lime, which is rare, the lime supply must be renewed periodically. On lime-deficient land it is usually necessary to spread lime about once in every five years at the rate of from $1\frac{1}{2}$ to 5 tonnes per hectare.

For the maintenance of maximum fertility grassland may require periodical dressings of a high phosphatic fertilizer. Old neglected pasture, where little or no phosphate has been used may require an initial dressing of 30 to 45 kg. per hectare, followed by annual maintenance dressings of 14 to 18 kg. phosphorus per hectare. Phosphates are slow-acting

5

fertilizers, which remain active for a long period; they stimulate particularly the growth of clovers, and often an abundant growth of wild white clovers may be called forth in fields where these had been thought to have been totally absent. Clovers in their turn attract nitrogen from the air and so stimulate root development in other grasses; they also retain moisture better than other grasses and so help to make a pasture more drought-resisting; finally, clovers are a most nutritious food, with a high protein content, and their presence enhances the feeding value of hay considerably.

Only potash-deficient land, mostly on sandy soils, may require dressings of high potash fertilizers initially 60 to 80 kg. per hectare, followed by an annual dressing of 15 kg. per hectare.

Whilst these dressings with lime and fertilizers correct the chemical balance of the soil, they do not add, at least not directly, to the so-desirable humus. In this respect there is nothing more effective than the spreading of farmyard and stable manure. This also is best applied in the autumn; freshly dunged meadows should be kept for haymaking, as they would not be fit to be grazed over until after that time.

With the approach of spring the time has come to prepare the pasture for the maximum possible production of luscious feed. To this end it is first necessary to plan the proposed use of every pasture on the farm; some will be required for early grazing, others for grazing a little later, and others still for making hay. In this respect it is important to note that the particular use whereto each field is put from year to year should be varied as much as is feasible, in accordance with the method known as rotational grazing. The importance thereof lies in the fact that the sward of a good pasture consists of a considerable variety of grasses and clovers, all of which it is important to preserve. But many of these varieties vary in their cycle of growth; some are earlier than others, some are deeper rooting, some are stronger growers, some will develop best in the shade of taller grasses, whereas others will not thrive when overshadowed or overcrowded. If therefore we always submit the same meadow to the same

6

treatment of early grazing, or late grazing, or cutting for hay, we shall influence, and gradually alter, the composition of the herbage. Certain grasses will be damaged and will finally succumb, to the detriment of the excellence of the sward as a whole which requires a healthy state of all grasses in the mixture.

The date on which pastures will be fit to graze, or to be cut for hay, can be advanced considerably, and the amount of fodder produced increased greatly at the same time, by giving them a top dressing of nitrogenous fertilizer in late winter, say February or March. A dressing of about 100 kg. per hectare of a 20:10:10 compound fertilizer will work as a strong stimulant to growth. Nitrogen is a quick-acting agent, but not in itself a complete fertilizer, and its effect is not lasting, although it does help the plant roots to assimilate more readily the proper plant foods that they require from the soil. Many horse owners prefer organic fertilizers which are released at a slower rate but there is no ground for the objection, sometimes raised, that nitrogen dressings exhaust the soil, provided it be realized that fertility must be maintained by periodical application of true agents of fertility, such as phosphates and stable manure.

Essential grassland cultivations consist in the main of harrowing and rolling. Both these operations should be carried out when the land is just nicely damp, without being sodden or muddy. And no land is ever fit for any kind of cultivation during or after a sharp frost; the ensuing thaw of the top layer renders it susceptible to severe damage if touched during such conditions.

Harrowing is quite the most essential, the cheapest and the most effective cultivation whereby to maintain grassland in prime condition and to keep on improving it. The chain harrows used for this purpose tear out moss, dead, old and tufty grasses, and so prevent or destroy the formation of any mat. The presence of such an old mat, as is so often found in parks and neglected pastures, prevents the access of air, sun and nitrogen to the rooting system, and stifles and destroys the finer grasses and prevents the development of clovers.

7

Harrowing opens the pores of the soil and helps the absorption of fertilizing agents.

It also spreads the animal droppings; these, when spread, form a useful manure and addition to the humus content, but when not so spread, and allowed to remain in heaps, they sour the particular piece of ground whereon they have fallen, cause rank and tufty growth, and spoil the pasture because animals will not graze over such places. Horses are particularly clean feeders, who will starve rather than graze off soiled ground. Moreover, droppings may contain parasites and worms, and, if left undisturbed, form breeding places wherein these pests thrive and multiply, and wherefrom in due course they will infect other animals. When these parasites are disturbed and spread about, sun and air will destroy them and keep the pastures clean and healthy. If public studs, where so many horses come and go, paid more attention to this point, there would not be so many cases of redworm infestation, which is so destructive of a horse's condition and injurious to his health.

The object of rolling is to put the surface down firmly; during the winter, with its sequence of frosts and thaws, through the absorption of much water, through poaching by stock turned out on wet land and kindred reasons, the topsoil goes hollow and loose; plant life cannot thrive under such conditions and consolidation of the topsoil is essential.

In order to derive the maximum benefit from our grazing, that operation itself will have to be managed with continuous regard to the condition of the pastures and to the health of the animals, two factors which are to a large extent interdependent.

Grass is most nutritious, its protein content at its highest, and it is most palatable during the growing stage; when it grows older, it becomes coarse and unpalatable, the animals will leave much of it, and the meadow will become tufty; where there are too many tufts for too long a period the finer grasses will disappear, and the herbage will in time deteriorate.

The ideal is to turn the animals in whilst the pasture is at its best, in the hope that they will graze it down beautifully evenly; in practice this ideal is difficult of achievement, and with horses alone it is quite impossible.

Horses are bad grazers, or rather they are difficult ones; they are very selective and will insist stubbornly on returning to the same patches over and over again, and will graze these down completely bare, whilst leaving adjoining patches, looking equally good to the human observer, almost or completely untouched; such left-over patches will soon become coarse, and under no circumstances will horses eat them.

Where horses alone are being grazed, the sward is bound to deteriorate and will go horse-sick, which is not only bad for the land but is also most detrimental to the health of the animals grazing on it. Undoubtedly the only satisfactory system is to inter-graze horses and cattle; that is even better for the land than grazing with cattle alone, because cattle too are selective grazers and leave certain places, although they do so in a far less marked degree than horses; in addition, it would seem that their preferences and those of the horse do in a sense complement each other, in that cattle will graze most of the grasses left by horses, and vice versa. To a large extent this is due to the fact that neither horses nor cattle will graze over their own droppings, but do not appear to be put off to the same extent when grazing over each other's droppings.

Another strong argument in favour of this inter-grazing has to do with the health of the animals. Both kinds pass quantities of worms and microbes of various kinds on to the land, where they are picked up by animals grazing after them; when picked up by animals of the same species they continue their life-cycle as parasites inside the animal's body, where they may in the process do a great deal of harm, as for instance in the case of redworm infestation in horses; but if picked up by animals of another species they cannot then continue their parasitical existence on the wrong type of host and are destroyed. In that way, inter-grazing helps to keep pastures free from infection and disease.

9

Horses graze with their teeth, they bite the sward, whilst cattle tear it off with their tongue; accordingly, horses are able to graze very close and they show a marked preference for the shorter grasses; cattle, on the other hand, cannot graze the shorter grasses and they go accordingly for the longer grass. Thus when cattle are taken out of a meadow because the grasses are becoming too short for them, there will be ample palatable and wholesome feed left to graze a bunch of horses for a few more days. And on the other hand, when horses are taken out because their favourite patches are becoming too bare, there will be plenty of long grass left whereon cattle will thrive until the field presents a more level appearance; if bulls are available on the place they can with great advantage be tethered on any such long-grassed patches, and they will make a beautiful job of levelling them and will manure them at the same time.

The best method of inter-grazing horses with cattle is in the manner indicated above, by grazing one after the other. Grazing them at the same time is not so effective and has other serious drawbacks; horses are too restless for cattle, they disturb them, they may chase them about, and they can easily cause a cow or heifer to slip her calf; on the other hand, cattle are not always safe with foals or yearlings; foals in particular are inquisitive little fellows and are fond of walking up and investigating; this may irritate some cow, and a bad case of horning may be the result.

The best time to turn animals into a fresh pasture is when the grass is from 4 to 6 inches long, and just begins to show a slight wave in a breeze; the grass is then at its very best in every respect. If we let the grass grow too long before turning out in it, it will lose some of its goodness and will no longer be quite so appetizing; also, much of it will get trodden down and be largely wasted, since animals will not eat such trodden-down stuff.

The best time to take the animals out is when the field has been nicely levelled down and has been grazed short, but not bare; bare fields retain little or no moisture, they will burn up quickly in hot and dry weather, and they will take a long time to recover; by leaving sufficient grass to retain a bit of moisture, we shall help towards a better and quicker recovery of the sward.

Each time we finish grazing a field, no matter how often this occurs during the season, and as soon as possible after we have closed it for its rest, we should have it harrowed thoroughly both ways. As already explained, this has the dual effect of a tonic to the field and of a health preserver to the stock.

That droppings, and particularly horse droppings, are a dangerous source of potential infection is so well known that some of the best-managed studs in the country make a point wherever possible of having all droppings, in every field, picked up every day and carted off the land; usually they have a man who keeps on going round all day with a tractor and trailer or a Terra-vac (automatic droppings collecting machine) to perform this task.

Where labour and expense are no object, that arrangement is of course admirable, but the same purpose can be achieved, at any rate nearly as well, by adopting the system of regular harrowing.

For the same reasons of health, and for the benefit of the pasture, horses should never be too thick on the land; in addition, too many horses together lead to restlessness, to jealousies, to kicking and biting, and sometimes to such vices as eating manes and tails; under such conditions horses do not thrive well. None the less, they should always have some company as lonely animals do not thrive either.

It is true that horses do exceedingly well on a large range, and half a dozen horses or so, roaming about in a hundred or two hundred acre park, will be very happy. But that is a condition hardly attainable on a stud, where a considerable number of animals may have to be kept, and where the necessity will exist to divide them up into carefully selected batches; in addition it will be necessary, on a stud, to ensure maximum grazing efficiency on the lines set out above.

For both these reasons, we shall need a number of comparatively small paddocks, as it is in that way alone that we shall be able to manage our grazing and to subdivide our animals.

With regard to the latter, the following is to be observed: mares and geldings should never be grazed together; whilst one gelding with several mares may not hurt, provided the mares are barren, the presence of another or more than one other gelding will at once lead to jealousy; similarly the presence of one mare between several geldings will upset their peacefulness; big strong horses and little weak ones do not make good company, more particularly so at times when hay or other feeds have to be carried and fed in the fields; breeding mares should not come into contact, not even across a fence, with geldings, since that may well cause them to slip their foals, particularly in the early stages of pregnancy; mares and foals should always have a paddock to themselves, and there should not be too many of them in the same paddock at that; mares heavily in foal should be kept quietly together and should not be mixed with barren mares or with younger horses that might cause them to gallop and run the risk of getting kicked; young horses should be separated from older horses, and from each other, according to sex, size and age as much as possible; young entires can be grazed with each other and with geldings up to about three years old, but not of course with fillies; colts may be unsafe with fillies from the time they are one year old; full-grown entires, if they are to be grazed at all, can only be turned out by themselves in a field wherefrom they cannot possibly get to other horses, and preferably wherefrom they cannot even see them; they must also be close handy for easy supervision.

On a stud of any size a good many paddocks will be required; for the reasons explained, it is advisable to have these paddocks of a handy size and not too big. In my opinion, fields of about 6 acres in extent are about ideal; they will accommodate about half a dozen horses, or mares with their foals, comfortably; they are big enough for them to stretch their legs when they feel like a gallop, and yet small enough to be grazed down in a fortnight or so, which is about the longest time that we want to keep animals in the same field;

12

they are of a handy size also for the owner's or stud-groom's daily round, for it goes without saying that on a properly run stud every animal must be seen, and inspected, at least once daily, and if the animals are scattered about over too much ground that task may become quite an arduous one. Also, in the case of studs taking in outside mares, there is occasional trouble with mares that are difficult to catch; to corner such animals in a six-acre field is one thing, but to try and catch them in a fifty-acre park is quite another.

It will be obvious that paddocks must be safe for horses, which is by no means as simple as it may sound, since horses possess a positive genius for getting into trouble, especially blood-horses. They are highly strung, impulsive and scatty, and very liable to injure themselves severely, and even to kill themselves, if their surroundings are not foolproof. If there be six inches of barbed wire, or even plain wire, anywhere near his field, a blood-horse is almost bound to find it, to get entangled in it, and maybe to blemish himself for life if he does no worse. As there is no sense in spending all the time, trouble and expense involved in rearing horses to no better purpose than getting them blemished, we must see to it that such things cannot happen.

Firstly, therefore, the fences must be safe. Any kind of wire, and especially barbed wire, of course, is out of the question. So-called park railings, made of iron, are also unsafe, excepting perhaps when high enough to be quite unjumpable; no fence much under 5 feet is unjumpable for a determined horse, or for one in a hurry, not even for a foal. A good hedge, where available, makes an excellent fence, again provided it be high enough and solid enough; any gaps must be repaired with posts and stout rails, and on no account with wire, since wire hidden in a hedge is more dangerous still than wire in the open. Where fences have to be constructed, the only proper and safe material to use are post and rails or Keepsafe fencing. Fences must be so laid out that acute corners are avoided; all corners must be rounded off; narrow triangular bits of land, such as occur sometimes, must be excluded from the paddock as horses are liable to get

13

injured in them. If there is a hedgerow or a row of trees to any side of the field, the fence will be built in front of them at a sufficient distance to prevent the horses from barking the trees. If any single trees or other obstacles stand out too far, so that the fence-line must be taken behind them, and yet not far enough to leave ample room for a bunch of horses to gallop freely in between such obstacle and the fence, it is necessary either to remove the obstacle or else to fence it in so that the narrow passage between the obstacle and the field fence be unusable. Without that precaution we would some day get a horse injured against such obstacle, possibly with a disjointed hip or with even more serious results. Each and every object against which the animals can possibly injure themselves, including the sides and corners of drinking-troughs, must be protected in a similar manner. Drinking-troughs must never be sited in any corner, where there would be no room for an animal to save himself if any of his companions should elect to turn spiteful; horses will seldom injure each other provided they always have sufficient room to get out of each other's way. For the same reason gates, where the animals are apt to congregate whilst waiting to be fetched in or for the arrival of their food, should never be in a corner, but should always have plenty of room on either side.

The field itself should be reasonably level, without holes, free from obstructions; no branches, fallen trees, bricks, stones or sharp objects of any kind should be allowed to lie about, and no implement of any kind should ever remain in any paddock used for grazing stock; or even in one intended to be so used, for its removal in time might be overlooked.

Gates should not be less than 12 and preferably 14 feet wide and 4 feet 6 inches high; narrow gates are a frequent cause of injury, and would not allow for the entry of tractors and farm machinery.

Fences and gates should be kept in good condition, broken rails replaced, and loose nails removed or knocked in; great care must be taken never to leave any nails about. It is preferable to keep all gates locked, and in cases where paddocks adjoin a road it is imperative to do so otherwise the horses may be stolen. Unfortunately, cases of trespassers or others forgetting to close gates are not rare, and apart from the annoyance of having to go and look for stock roaming about the countryside or on the roads, there is in such cases always a grave risk of accidents.

The presence of too many trees is no advantage to a paddock; trees are very competitive in absorbing fertility from the soil; on account thereof and also on account of shade, the grass is always of an inferior quality in the neighbourhood of trees. Yet it is almost essential to have some trees, whereunder the animals may find shelter from sun and rain, but particularly from the former. At the same time, if we value our trees and wish to retain them, we will do well to remember that horses are destructive of them; they delight in tearing off the bark, which they eat, presumably for the sake of the minerals which that contains, or possibly only because they like doing it, or out of mischief or ennui. In so doing they can manage to inflict the tree's death sentence amazingly quickly, even though the injured tree may well linger on for several years. Horses seem to injure almost every kind of tree in that manner, with the exception of English oak, which I have never known them to touch. The only effective way of preventing such injury is to erect tree-guards where required.

Some people imagine that the placing of blocks of rock-salt in every meadow, which the horses may lick in their craving for minerals, will induce them to leave the trees alone. It will do nothing of the kind, but even so it remains very necessary to give the horses free access to mineral salt, and from that point of view this distribution of lumps of rock salt or salt lick blocks to every paddock is thoroughly recommended.

In addition to the presence of some trees, it is an advantage to have a stout hedge, a clump of trees or some other such protection, behind which the animals may find some shelter against piercing northerly and easterly winds. But it is no advantage to have thick overgrown hedges, full of thorns and brambles, or to have too much bushy growths near our

15

paddocks; such places are nothing but rabbit-warrens and breeding places for flies and other pests. Rabbits are very detrimental to the land, and no grass will grow where they have soured it. For the same reasons, I do not like the close proximity of too many woodlands. Apart from the dangerous holes they dig, which must be filled in each day if broken legs are to be avoided, there is the old saying that, when grazing, 4 rabbits equal one sheep and 4 sheep equal one cow.

The ideal paddock should lie on a gently southerly slope, be open to free access of sun and air and yet provided with some shelter against an excess of sun and against the coldest winds.

Some studs lay great store by the erection of shelters in their paddocks; these consist normally of a building enclosed to three sides, open towards the south, about 12 feet high to eaves and covered in with a thatched or tiled roof; corrugated iron is too highly conductive of heat and quite unsuitable as a roofing material for any stock building, but substitute asbestos sheets may be used. Although such shelters are hardly essential, they are certainly an advantage, especially for young stock during the hottest period of the summer, and also during the winter. Provided always that the use of such paddocks can be limited to a small number of horses only; if too many animals are able to crowd into such places they become a source of danger, as all confined spaces do.

The final requirement is the provision of an ample supply of fresh and wholesome drinking water to every paddock; since stagnant pools or ponds are neither wholesome nor fresh, but are on the contrary nothing but dangerous sources of infection, they will quite obviously not do; any such pools as may be present will have to be fenced off. Any water from a running stream is also not suitable in most areas, as it is often contaminated with nitrogen fertilizers and chemical sprays washed off the fields. Near industrial areas it may be contaminated from time to time with other chemicals and waste. It will therefore be necessary to take a piped supply to a drinking-trough in each field. To save trouble, these troughs should be fitted with a ball-valve and be self-filling; even so, it should be made a matter of routine to check the water daily, as the self-filling arrangement does go out of order sometimes.

The water provided by such drinking-troughs will remain clean and healthy as long as they are in regular daily use; but as soon as the animals are taken away and in warm weather the water in them will become stagnant and foul, and unsuitable for further use; here again it should be made a matter of strict routine to clean out every drinking-trough thoroughly each week during the summer and upon each occasion when animals are taken into a fresh pasture.

Water pipes leading to such troughs should be protected as well as possible against frost; even then, they, and the tanks themselves, may become frozen in cold weather. It is essential not to forget to inspect all drinking places at least twice a day during sharp, frosty weather, to remove any accumulated ice from the troughs, and to thaw out the supply pipes, if necessary, with a blowlamp.

Hay constitutes the bulk of our winter feeding, for which it is quite irreplaceable and completely indispensable. At a pinch we can do without concentrates, but under no circumstances can we do without hay. Since hay is such an important crop, it is obvious that no effort should be spared to make as much hay as possible and of the best attainable quality. In point of fact, quality is more important than bulk. Quality depends in the first instance upon the botanical composition of the herbage which, in the case of meadow hay, should contain a good admixture of rye grass, rough-stalked meadow grass and clover; the latter in particular enriches the hay in protein and lime, and so in feeding value. Meadows containing largely fibrous bent, crested dogstail and other inferior grasses, are incapable of producing anything but poor hay.

The following table will show the immense difference in feeding value between prime and poor meadow hay:

17

	Dry Matter %	Crude Protein %	Digestible Protein %	Digestible Energy MJ/kg
Best Meadow Hay	85.0	13.5	9.2	10.0
Poor Meadow Hay	85.0	4.0	2.0	6.0

Accordingly the feeding value of best quality hay is something like three times superior to that of poor quality. But however valuable and instructive a chemical analysis of foodstuffs may be, animals, like humans, need something more than that; they require their food to be tasty and wholesome.

This brings us to the second essential whereon the quality of hay depends, namely that it must be thoroughly well made. Actually that is, in a sense, more important still than the composition of the herbage, since poorly-got hay will be of little value no matter how good the grasses that went into it.

In the first place, hay must be cut when the grasses are at their best as regards protein content and wholesomeness; that is when they are still young, just as they enter upon the flowering stage and before they have run to seed. In a normal season that stage will be reached towards the last week in May or the beginning of June; at that period the grasses are still in their growing stage and, if left, will of course make more bulk; but the greater bulk will only be obtained at the cost of reduced quality; the grasses will run to seed and the greater bulk will consist largely of an increase in fibre, at the cost of a decrease in protein.

In the second place, as everyone knows, hay must be made when the sun shines; but it is no advantage to have it too hot, a combination of sun and breeze being about ideal. And though everyone knows that rain is liable to spoil hay, not everyone seems to realize that too much exposure to the heat of the sun may do almost as much damage; that will bleach the hay, when it assumes the colour of straw; in so doing its feeding value and taste will be reduced to much the same inferior quality. Good hay should preserve its green colour

18

and with it the aroma or nose that is so characteristic of the first-class article.

Hay is fit to be carried when the sap has dried out, it will have lost the deep green colour of sappy grass and will have assumed instead the blue-green colour of first-class hay; it will feel dry and warm to the touch and will rattle when handled with a fork. As long as it contains too many traces of sappy grass, feels clammy to the touch and fails to rattle, it is still unfit. The art of haymaking consists of baling it as soon as it is just fit, neither an hour earlier nor later, which requires a nice judgment.

Though the conception that meadow hay is an inferior food for horses is quite erroneous as far as first-class meadow hay is concerned, there is yet no doubt that horses appear to prefer so-called hard or seeds hay. In this respect it is a definite advantage to have some land down to seeds; the most usual are pure clover, clover mixture, the old favourite sainfoin, or lucerne.

The following illustrates the comparative feeding value of different classes of hay in protein content:

Poor Meadow Hay	4.8%
Fair Meadow Hay	7.6%
Best Meadow Hay	13.5%
Clover Mixture Hay	12.0%
Clover Hay	13.0-18.4%
Sainfoin Hay	13.9%
Lucerne Hay	17-19.3%

Clover mixture hay is made from a mixture of red clover and rye grass; clover hay from pure red or crimson clovers. Sainfoin hay is not only very rich in protein, but also in lime; lucerne hay is by far the richest of all, since its protein content may exceed 10% and its lime content be as high as 1.43% as compared with only 0.33% for some grass hays, so that it is really unrivalled in feeding value.

Lucerne, known in America as Alfalfa, is such a precious, abundant and remunerative crop that no stud farm ought to

be without it; it may be cut from three to five times in a season; it will produce two hay crops and an abundance of green fodder as well; it is absolutely drought-resisting and can be relied upon to go on producing luscious green fodder even when all other forms of pasture have long since burned up. With its deep rooting system it is a soil improver without equal. It will not thrive in wet areas with a rainfall in excess of 88 cm. Whilst it is not an easy plant to establish, requiring great care and a very clean seed-bed, it will last a long time; anything from 7 to even 15 years under the most favourable conditions of a warm dry soil. Dried lucerne is also available for feeding to horses as an 8 mm. cube or as dried lucerne chop. This is particularly valuable during the winter months when fresh green forage is scarce.

CHAPTER THREE

The Buildings

General Observations—Layout—Cottages, Loose-boxes, Building Materials—Mangers, Hayracks, Drinking-bowls, Light, Slings, Tackroom, Dutch Barn, Chaff-cutter, Feedhouse, Corn-store, Manure-pit, Toolshed, Cartshed, Ladders, Forge, Carpenter's Shop—Road Surfaces—Plans and Specifications.

IT IS an accepted truism that there can never be too many buildings on a farm, and that applies most certainly to a stud farm. But there can be a vast deal of difference between one lot of buildings and another, and whereas a well-arranged and properly adapted set of buildings will be of tremendous value, poor buildings may be of almost equally great detriment to the efficient running of a stud.

Naturally, where money is no object, it is easy enough to plan and erect the perfect type of stud farm, provided always that the planner knows what he is about. At the same time, it if often possible to obtain almost equally good results, at a fraction of the expense, by the intelligent adaptation of existing accommodation. Quite excellent results can be obtained, for instance, by the conversion of old barns and other existing farm buildings, and more often than not the final result will be found to be most pleasing from an artistic as well as from a utilitarian point of view.

But whatever method or variety of methods are adopted, there are a few golden rules that apply to every kind of building and that one will do well to remember.

In these times, all building is expensive, and so therefore are extensive building repairs. The greatest enemy of buildings is damp. Roofs must be kept in first-class order and all rainwater intercepted by ample gutters and downpipes. If gutters and downpipes are allowed to get obstructed with leaves or birds' nests, or other rubbish, they will overflow and the water will rush down some part of the wall and will most certainly ruin it. Downpipes must be connected with gullies and drains, for if the water is allowed to gush down on to the ground it will find its way into the foundations and in course of time cause a lot of damage. Roofs, concrete yards and similar areas collect an enormous amount of rainwater, and if we wish to economize on future repair bills it is essential to get rid of all this water and to take it away from the buildings. For the same reason, all buildings should be kept above the level of the surrounding land and any new erection be provided with a damp-course.

All woodwork and all ironwork should be protected from the weather by painting, creosoting or tarring; ironwork may be treated cheaply and effectively with bitumastic paint. Generally, all small defects should be put right as soon as noticed, for if that is not done small defects will in course of time develop into big ones; remember that a stitch in time saves nine.

All building furniture used in connection with stud buildings, such as hinges, locks, bolts, door handles and the like should be as strong as possible; and certainly be of considerably greater strength than similar fittings used for domestic buildings; they will be subjected to very hard wear, and inferior quality fittings will cause nothing but trouble and expense. All doors that have to be left open frequently, such as doors of boxes, should be fitted with hooks by means of which they can be fixed easily in the open position; otherwise they will be found to be banging themselves to pieces whenever there is a strong wind blowing. Any material, brick, timber or otherwise, used for the construction, repair or adaptation of stud buildings should be of good quality and of ample strength. Horses are strong animals and flimsy materials will soon prove to have been false economy; tops of doors or other similar places that horses are liable to bite, should be constructed of oak or, failing that, be covered with a strip of metal. If these and similar points are borne in mind, the saving on the maintenance bill will be considerable.

22

The layout of the stud buildings is of great importance. The first thing to consider is the need for drainage, so that both surface water and foul drainage can be got away reasonably easily; no buildings should be sited in such a way that they risk flooding in cases of heavy storms or melting snow. The second thing to make sure of is that the layout is as compact as possible. The care of animals involves a lot of moving about, and unless all requirements are grouped together in a handy way, time will be lost or, what is worse, work will remain undone.

The stud-groom's house and any staff cottages should be close to the stud buildings, say within a couple of hundred yards at the outside. This is absolutely imperative, since the care of livestock requires close and almost constant supervision; there may be sick animals or foalings to be attended to, necessitating visits during the night; some horse may get out, or get cast, or get into trouble in some other way, and if the man's house is near enough, he is certain to hear any such commotion and to attend to it.

Whilst on the subject of cottages, it may be as well to mention that a good man will expect a reasonable standard of comfort, and that central heating and a hot water supply are essentials today.

The arrangement of the stud buildings proper will depend, of course, upon the size of the stud and the number of boxes likely to be in use. But the following observations will be found to apply in a general way.

Boxes should be sited with a view to receiving plenty of sun and air, and yet be protected from north-westerly to north-easterly winds; a southerly to south-westerly aspect is to be preferred, but westerly and easterly aspects need not be avoided provided there is some wind protection, particularly from the east.

The groom's workroom, tack- or saddle-room should be in a central position and adjoining or very near to any boxes that it is proposed to use for foaling; the feedhouse must also be close at hand. In addition, there will have to be a corn-store (unless incorporated with the feedhouse) and a hay and

23

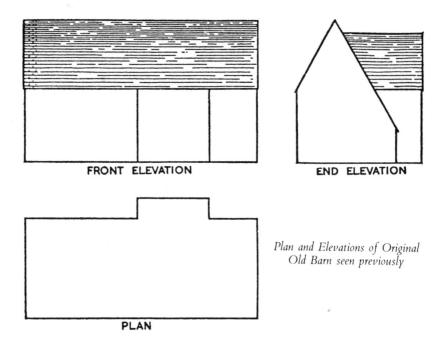

FRONT ELEVATION END ELEVATION

*Plan and Elevations of Original
Old Barn seen previously*

PLAN

FIGURE 1. AN INTERESTING CONVERSION

straw barn with chaff-cutting shed attached; hay and straw
are bulky commodities that have to be moved several times
daily to every box, so that it is essential to have this barn in an
accessible place. One of the most laborious jobs on any stud is
the mucking-out of boxes, and it is therefore essential to have
the manure pit in a handy place as well; but since manure
heaps are not exactly pleasant to look on, they should none
the less be out of sight, or surrounded by a wall or a hedge. A
toolshed, for wheelbarrows, forks and similar equipment
may be arranged near the manure pit. It is to be borne in mind
that the ammonia contained in manure is damaging to
buildings, and for that and sanitary reasons as well, the pit
should not be too close to any building. Most studs will
probably also require a tractor and implement shed, possibly

24

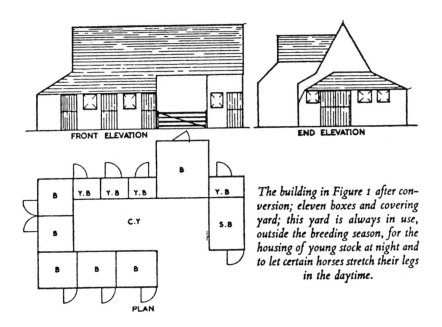

FRONT ELEVATION

END ELEVATION

PLAN

The building in Figure 1 after conversion; eleven boxes and covering yard; this yard is always in use, outside the breeding season, for the housing of young stock at night and to let certain horses stretch their legs in the daytime.

FIGURE 2. AN INTERESTING CONVERSION

with enough room to house a horsebox or a trailer; such a building will not be in constant use and may well be at some distance from the stud buildings proper; it may be convenient to erect it in some place where it will serve as a wind-break. The same remarks apply to a garage that may be needed. Large establishments will have a forge.

If stallions are to be kept, the question of stallion boxes and of a covering-yard will arise, and on big studs the possession of a covered riding school is often considered indispensable.

As I have pointed out already, it is difficult to have too many buildings, and the possession of one or more covered yards suitable for wintering yearlings or other young stock is always an advantage.

Since horses drink a great deal of water, it is essential to have water laid on to almost every building, and it is equally as essential, in my opinion, to have electric light in every building.

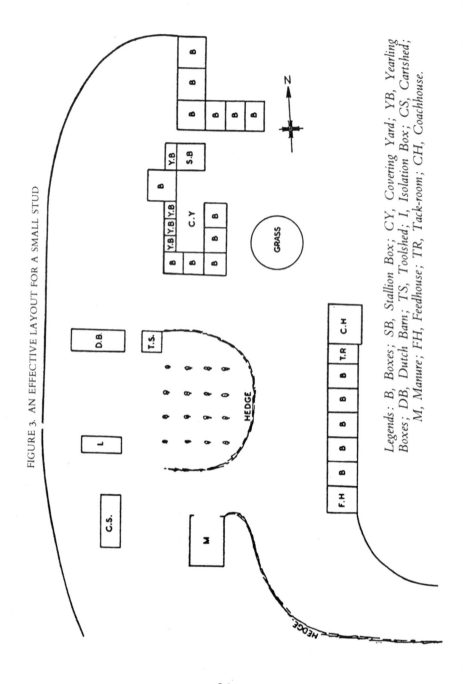

FIGURE 3. AN EFFECTIVE LAYOUT FOR A SMALL STUD

Legends: B, Boxes; SB, Stallion Box; CY, Covering Yard; YB, Yearling Boxes; DB, Dutch Barn; TS, Toolshed; I, Isolation Box; CS, Cartshed; M, Manure; FH, Feedhouse; TR, Tack-room; CH, Coachhouse.

The final requirement of a good layout is easy access from the boxes to at least two or three small paddocks for the use of brood-mares and foals.

After this survey of general requirements, we may now consider the main points that arise with respect to particular buildings. It will be appreciated that the need of saving labour must be considered at all times, since labour is today both scarce and expensive.

As stalls are absolutely unsuitable for stud work, it is assumed that loose-boxes only will be used. The number of boxes will depend upon the number of horses to be kept, with the essential proviso that there must always be some spare boxes available as one can never tell what contingencies may arise from sickness, accident or other causes. The ideal is that one should be able, at a pinch, to house all one's stock at one and the same time and still have a little room to spare.

The size of a box may vary from about 3.5 x 4 m to about 4 x 5.5 m; the so-called portable loose-boxes are often made in the former size, but this is much too small for stud purposes, excepting perhaps for pony breeds. It is too small even for Arab mares who only measure about 14 hands 2 inches on an average. In my opinion, 4 x 5.5 m is suitable for any type of horse and for any type of mare and foal. On some studs one or two much larger boxes are kept specially as foaling boxes, but I do not consider this necessary, as a box of 4 x 5.5 m is roomy enough for foaling purposes. And at any rate, for reasons to which I shall refer later under the subject of 'foaling', I do not advise the use of any boxes kept specially for foaling.

More important perhaps than the width and length of our boxes is the amount of air which they contain, and therefore the height. That is the dimension most frequently sinned against, no doubt for reasons of economy. Whereas portable loose-boxes are often seen that measure only about 2.6 m to eaves, and maybe 3.8 or 4 m to ridge, I consider that the free height in a stable should not be under 4 m anywhere; which is

27

one of the reasons why old barns often make such excellent boxes.

Doors should be 2.6 m high, with the bottom door 1.5 m high, and not less than 1.3 m wide; narrow doors are very dangerous. In addition to the door, each box should have a window for the purpose of light as well as of ventilation; the provision of further draughtproof ventilation in the roof is an advantage. Horses need plenty of fresh air, but free from draughts.

Perhaps the most important constructional part of a box is the floor; which must be hard and impervious, non-slippery, and any liquid must drain away. The so-called Stable Paving Bricks, usually made in Staffordshire, and which are obtainable in a number of patterns, are undoubtedly the best flooring material, but they are rather expensive. Almost equally good results may be obtained with a concrete floor at considerably less expense. In old-fashioned stables, floors are usually drained into a drain-trap situated inside the box and in the centre of the floor. That system is most unsatisfactory, since it will constantly caused blocked drains, and also because it is liable to cause an unhealthy smell and gas in the stable. Open drains, with movable covers, that lead to the outside are much better, but are really also unnecessary.

In my opinion, the best plan is to give the floor a very gentle slope of, say, 2.5 cm in 1.6 m towards the doorway, with a few shallow grooves about 4 cm wide by 1 cm deep to the lowest point; these grooves will be quite sufficient to take the liquids away and to keep the bed dry, and they will run out into an open channel just outside the door and from there into a drain-trap outside the building. In that position the channel can be seen and will be kept clean, and the drain-trap will not get obstructed.

The most usual materials for the construction of boxes are bricks and timber. Brick is assumed to be the more durable of the two, though it must be admitted that wooden buildings, provided they have been constructed properly and of the right kind and size of seasoned timber, will last a very long time; some of our wooden houses and timber barns have

*The most satisfactory system is to intergraze horses and cattle.
An example of a nicely sheltered paddock.*

Foal's first day out in a paddock safely fenced with post and rails.

Main stable yard, with abundant space, light and air. Right foreground: range of brick stabling, feedhouse, tack-room and coach-house. Left background: range of weather-boarded stabling built round a converted barn.

An interesting old tithe barn before conversion, south-west aspect.

Same aspect after conversion. The eaves have been lifted to accommodate yearling boxes; the dormer doorway has been made into a large and airy box.

Same building, south-east aspect, showing how five full-size loose boxes have been added to the south and east fronts.

Same building, east aspect showing stallion box on right, and gate entrance to covering yard, also used for housing young stock in winter.

'Gainsborough', winner of the triple crown.

'Bahram', winner of the triple crown.

The thoroughbred stallion 'Big Game'. The English Thoroughbred is capable of exerting a very potent influence for the improvement of inferior breeds.

Thoroughbred stallion 'Duncan Gray' in the stallion exercising yard.

Champion Arab stallion 'Raktha', a Kehilan Rodan, by 'Naseem-Razina'.

Modern Arab stallion by 'Sky Crusader'.

Champion Arab filly 'Grey Royal', a Kehilan Dajanieh by 'Raktha-Sharima'.

Two-year-old Arab colt 'Dargee'. The Arab carries, compressed in his small body, the greatest array of equine qualities.

stood up for two or three hundred years. Apart from attention to weather protection in the form of tar or paint, timber buildings do not require much maintenance, and repairs are usually easy and cheap. Walls of brick-built boxes should not be less than 23 cm thick, and timber buildings ought to be lined inside; wooden boxes must of course be erected on a brick or concrete foundation wall.

The best type of roof would be close-boarded and tiled, to ensure not only sound weather protection but also substantial insulation against heat and cold. Roofing felt is a substitute of poor appearance and does not last. Corrugated-iron sheets are quite unsuitable as a roofing material for any livestock buildings, as their insulating qualities against heat and cold are almost non-existent. They are, moreover, unsightly and out of place on a stud farm, excepting for use on Dutch barns; even then they should be kept well painted with bitumastic paint, both for the sake of rust protection and for appearance.

Every box should be provided with a manger; in boxes used for mares and foals the mangers should be double, or at any rate of sufficiently ample size to allow a mare and foal to feed together in comfort; corner mangers, that are quite suitable for stallion- or hunter-boxes, are no good for brood-mare boxes, since they do not allow the foal sufficient room. Brood mares should have the manger fitted in the centre of the short wall, away from the door. Mangers must be substantial and easy to clean; whilst there is nothing to beat porcelain or glazed earthenware, porcelain-enamelled cast-iron mangers are also suitable. Some of my very knowledgeable friends swear by having ground-mangers, which means fixing them at about 15 cm above ground level. Whilst they are no doubt correct in asserting that it is natural for a horse to feed off the ground, I must admit that they have never been able to convince me; I have an open mind on the matter. Meanwhile, I make a practice of having my mangers fixed at about 1 metre above ground level. Hay-racks are not required in stud boxes; the old-fashioned type that used to be fixed high up against the wall is not recommended because horses feeding from them are liable to be troubled with dust in the eyes; the more modern low-down type, usually fixed alongside the manger, quite useful though they are in themselves, are not recommended for stud purposes, because foals may get entangled in them. In fact, the less obstructions there are in a box the better, and whatever obstructions are unavoidable, a manger, for instance, must be built-in and protected in such a manner that foals cannot injure themselves against them.

Personally, I like to have an automatic drinking-bowl in every box, firstly because my horses are in that way assured of a never-failing supply of fresh drinking water whenever they require it, and I believe that to be essential to their well-being, and, secondly, because it saves the very considerable amount of labour involved in carrying buckets to every box several times a day. And even though buckets may be filled with religious precision every night at evening stables, horses will still be without water towards the middle of the night, especially in hot weather; brood mares, in particular, require quite a lot of water, and without it their milk yield will suffer.

An electric-light point in each box is almost essential, and certainly desirable in all buildings. Switches should be placed outside the horses' reach, as otherwise some of them become adept at turning the light on; they may be placed outside, provided the watertight variety is chosen. Every box should be provided with a tying-up ring, anchored solidly in one of the walls.

Window panes must of course be protected by iron bars no more than 8 cm. apart, or by weldmesh panels. Doors must be locked from the outside, one bolt to the top door and two bolts to the bottom door; the second bolt should be fixed at about 15 cm. from ground level, and the so-called kick-over variety are the most convenient. This second bolt at the bottom is essential, as some horses are able to open almost any kind of bolt that is within their reach.

Some studs have extra large boxes for their stallions; though these are no doubt convenient, they are not essential, as long as sufficient outside exercise can be given. Others have their stallion-boxes connected with a large open yard or with a small paddock, wherein the stallion can be turned out to exercise himself, and this arrangement is effective and time-saving. Other studs still have their foaling-boxes connected to either covered or open yards, and that arrangement saves the necessity of turning mares out during the winter and is also useful for mares and foals during the early part of the foaling season. But all these arrangements require a lot of room and they may not suit everybody's purse, and, of course, they are none of them essential.

What is essential for every stud of any size is the provision of one or two isolation boxes for the treatment of sick animals; as certain affections are contagious, such boxes should be at some distance from the regular yard; one of these boxes ought to be fitted with a strong overhead beam for the purpose of accommodating an animal in slings. This beam must be placed so that an animal supported in slings will be able to reach his manger and his water-bowl easily. these two being, in this case, arranged side by side.

Slings are suspended from the ceiling by means of an endless chain and a multiplying block, so that they can be adjusted easily to the exact level required. The purpose of slings is to give rest to an animal that must be kept in the standing position for a long time, possibly for several weeks. They should be so adjusted that the animal can rest his weight on them whenever he wants to, but the horse should on no account hang in them, as that would rapidly cause acute pain and discomfort.

The groom's workroom should be well lighted and be provided with heating and with hot and cold running water. It is best to have one workroom and a separate room for the storage of clean tack, blankets and rugs. A suitable place should be provided for drying wet clothes, stable rubbers and similar articles. On large studs it may be necessary to provide also a messroom for the stud hands. There should be ample cupboard room for the storage of blankets, small tools and medicines.

31

The possession of a Dutch barn for the storage of hay and straw is of such an enormous advantage as to make it almost indispensable. Where hay is grown at home, and not purchased in small quantities as required, such a barn should be big enough to accommodate one whole year's supply. Allowing for the times spent on good grazing, when the consumption of hay will be very low, I think it is fairly safe to estimate the annual consumption at from $1\frac{1}{2}$ to 2 tonnes of hay per horse, which is equivalent to from 15 to 20 cubic metres. As loose hay, before settling, will take up half as much room again, somewhere around 30 cubic metres of space should be allowed per head. Assuming the height to eaves to be the usual one of 6 metres, then a floor space of 2 x $2\frac{1}{2}$ metres will hold all that is required for one horse. Thus a building of 10 x 5 metres will be sufficient for 10 animals, one of 20 x 5 metres for 20 head, one of 15 x 8 metres for 20 horses, one of 20 x 8 metres for 32, and so on. If the hay can be baled before stacking, more than half the room can be saved. Naturally, some additional room will be needed for storage of bedding straw; as this can usually be got in smaller quantities, a bay of about 5 x 8 metres by 6 metres high will look after a dozen horses, and if the straw is baled the capacity will become accordingly bigger.

The usual type of Dutch barn is a rudimentary sort of building, consisting of a number of uprights with a roof on top; if the building is open to all sides, too much rain will blow in and spoil much of the forage; on that account it is essential to cover in the whole of the weatherside, and the back and remaining side to at least halfway down from the eaves. The open front, which should not face in the direction of the prevailing wettest winds (S.W. to N.W.) can with advantage be covered in to about 1 metre below eaves, but no lower, as that would interfere with loading operations. In that way there will be sufficient protection from wet and yet plenty of circulation of fresh air, which is so essential with newly-made hay.

But even though the protection from rain, and even from

driving rain, will be quite sufficient, yet a certain amount of water is bound to be blown in at times. This will do absolutely no harm, provided the levels of the barn floor have been made right; the floor, which should be of concrete, must be highest in the centre of the building and from there have a slight fall to all outsides; thus any water finding its way into the building will drain away without percolating underneath the entire stack as it might do if, for instance, the entire fall were from the front to the back of the building only.

The front of the barn will have to be sited in such a way that there is plenty of room for an elevator to be placed in position in front of each bay, for loaded lorries to be manoeuvred in front of the elevator, and for traffic to proceed on its way or to be turned about. That requires a yard of about 17 metres wide in front of the barn; such a yard should be surfaced with tarmac or concrete, and the floor level of the barn kept from 10 to 15 m above the level of the outside yard.

Barns constructed of metal, whether galvanised or not, should be painted all over with bitumastic paint.

A lean-to roof can be arranged conveniently against one of the sides or to the back of the barn for the purpose of housing a chaff-cutter. As chaff-cutting is a dusty job, it is not advisable to have this machine in an enclosed building, and operators should always wear a mask.

The feedhouse, which may adjoin the boxes and must be in close proximity to them needs to have adequate access: for like horse transporters, delivery vehicles get larger every year.

If corn is to be stored in bulk it is both difficult and expensive to take deliveries under 10 tonnes. Therefore an outside round self-emptying bin of 12 tonnes capacity is recommended. This is because the bushel weight (volume: weight) of oats varies from one load to the next and season to season. A bin holding 10 tonnes in one delivery may be too small for the next delivery — so a margin of safety is indicated.

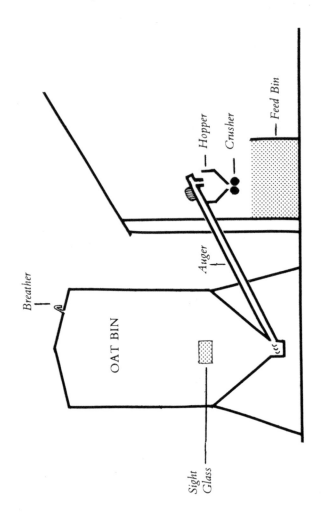

FIGURE 4. SELF-EMPTYING OAT BIN AND MOST CONVENIENT
ARRANGEMENT FOR CRUSHING THE OATS. THE OATS ARE
DELIVERED IN A BULK BLOWING LORRY SO ADEQUATE ACCESS
IS IMPORTANT

The bin should be of metal construction, fitted with a sight glass at the 1 tonne level and a breather at the top. An auger should be fitted to the bottom of the bin passing through the feed house wall. This connects with a small holding hopper over the crusher, which in turn should be sited directly over a metal feed bin. Since processing oats creates dust, it is recommended that a dust extractor unit should be fitted to the crusher. The purpose of mounting the crusher directly over the feed bin minimises the distance the grain is moved after crushing. Everytime oats are moved after being crushed, dust is created and losses through spillage may occur.

Ideally the small hopper over the crusher should be fitted with two pressure switches: one at the top and one at the bottom. The top switch will stop the auger when the hopper is full; the bottom one will switch off the crusher when this is empty. This is a simple system which allows one to feed freshly crushed oats every day. Like newly ground coffee, freshly ground oats have a characteristic aroma which soon fades. The machinery is driven by electricity available in both single and 3 phase.

Other items required in the feed house include:

1. Electric boiler/steamer;
2. Scales — for check weighing;
3. Feed barrow;
4. Wooden pallets — to keep bagged feed off the floor.

Useful dimensions of floor space:
1 tonne of cubes in bags stacked 8 high: 6' x 4' (2 m x 1.3 m) ½ tonne of bran or flaked grain stacked 6 high: 6 ' x 4' (2 m x 1.3 m)

When purchasing whole oats in bulk the following points should be noted: The moisture content must not exceed 15.5%. The colour should look bright. They should be free from admixtures i.e. other grains such as barley and especially wheat, also weed seeds, dust or moulds. Moulds should not be confused with weathering which occurs after bad conditions at harvest time. There should also be an absence of insects which are commonly found in stored grain, such as the Saw-Toothed Grain Beetle, and Lesser Grain Borer. An infestation of say, Saw-Toothed Beetle will generate heat and humidity from their bodily functions, causing the grain to develop a bad smell. The grain is then unpalatable to horses as well as having a reduced food value since the kernel has been eaten by the beetle. The breather at the top of the bin allows cold air to be blown through the bin when necessary.

The correct and convenient siting of the manure-pit is of much importance for the reasons already stated. On a well-run stud, where horses are given ample bedding, which is almost as important as ample feeding, the amount of manure produced is very considerable. It is wise, therefore, not to under-estimate the size of the pit required. The pit should consist of a concrete floor surrounded by brick or concrete walls, about 1 to 1.3 m. high; the floor should be constructed with a slight fall to connect with a drain that will take away the moisture that will seep through to the bottom. According to the best practice, the manure-pit should be covered over with a simple form of roof to keep off the rain. It is advisable to have a double pit; when the first pit is full, the contents are turned over into the second pit; through this turning process the manure will make, and go down in volume at the same time, so that the first pit can usually be emptied several times into the second pit before the latter is full as well; when that occurs the contents can be carted away to the fields. There must be sufficient room, therefore, to manipulate tractors to and from the manure heap. Many large studs have an ongoing arrangement with a firm of manure dealers to clear their manure heap on a regular basis. The quantity of manure collected by the end of a busy season would be too great to spread on the paddocks.

The toolshed need be no other than an unpretentious small building, yet one that has sufficient room to store the usual mucking-out tools out of sight in an orderly manner. The implement-shed is usually left open to the front and closed in at the back and sides; it is generally economical to erect this building, which is not intended to be more than a weather protection, in weatherboarding. Implements left out in the rain deteriorate quickly, and it is sound economy, therefore, to construct this shed of sufficient size to accommodate all the implements we have. If a horsebox or trailer is to be housed as well, it is necessary to allow for sufficient height in the building. Any type of estate invariably requires a set of ladders, but these will rot away quickly if allowed to lie about in wet grass; a lean-too roof against the back of the implement-shed is often the most convenient place for the storage of ladders.

If a forge is maintained, as is done frequently on the larger studs, this had better be sited at some distance from the stables on account of the noisy nature of the work. It can be housed in quite a simple building, which must however be well lighted and well ventilated; the machinery required is simple enough and need consist of nothing more than the usual farriers' tools; the addition of a good drill and of a small welding plant are, however, recommended, as well as the fire.

In my opinion, a small estate carpentry shop is almost essential on a stud farm, where there are bound to be constant repairs to fences, gates and boxes, and where work can always be found for a good all-round carpenter-handyman.

The surfacing of yards, passages and drives deserves mentioning. Where expense is no object, the best effect, both from the point of view of durability and of looks, is obtained with ribbed paving bricks laid on concrete; the buff-coloured variety look particularly well. However, the cost may be found rather prohibitive if large areas are to be covered. Sometimes a path, up to about 2.6 m. wide, is laid in this material in front of the boxes only. Equally good results, from the point of view of utility and cleanliness, may be obtained at considerably less expense, with concrete paths in front of the boxes. Asphalt, which is sometimes used, is hardly suitable, because it is at once too slippery and too soft. Though concrete can be recommended for pathways in front of the boxes, I am not fond of this material for surfacing yards and roadways on a stud farm; it is altogether too hard, too rough, and too dangerous a surface whereon to lead high-couraged horses. A better one, almost equally clean and durable, can be obtained with one of the many types of tarmac surfaces that are normally available. Gravel, well rolled in, is a simple and pleasant surface, which does, however, require a fair amount of upkeep and is apt to be somewhat muddy in wet weather. Cinders or ashes are useful for surfacing outside farm-tracks, but they are a dirty and dusty material, out of place anywhere near boxes or stable-yard.

Wherever new tracks or yards have to be laid down, it is necessary to dig out the topsoil to a depth of at least 15 cm, and to begin with a foundation of hardcore, rubble or coarse gravel of at least 15 cm thick; this should be well rolled in and left to settle for a while before constructing the top surface.

And, of course, careful attention must be paid to the inevitable problem of draining the surface water.

Any building problem, however small and unimportant it may appear to be, is liable to present peculiar difficulties of its own which may, if not realized in time, become the cause of very considerable trouble. It is always advisable, therefore, to employ a qualified architect or surveyor to draw up plans and specifications for such work.

I have illustrated the subject matter of this chapter mainly with plans and pictures of my own buildings, which I have thought to be interesting, because almost all of them were achieved by adapting, or converting, or adding to existing buildings. Although I am far from pretending that the examples shown are perfect, and fit to serve as a model, I am satisfied from experience that they constitute an effective and workable layout, and present a reasonably pleasant and attractive appearance. They may perhaps be of some service, therefore, to readers intent upon formulating their own particular ideas.

CHAPTER FOUR

Horses of Britain: Principal Breeds & Types

The Thoroughbred—Warm-blooded and Cold-blooded Breeds—The Arab, Anglo-Arab and Part-bred Arab—The Hunter—The Hack—The Cob—Docking—The Hackney—The Cleveland Bay—Ponies—The Polo Pony—The Native Ponies—Cold-blooded Horses: The Shire, The Clydesdale, The Suffolk, The Percheron.

THE English Thoroughbred is without a doubt Britain's and the world's premier breed of horses. Although bred primarily for racing, a couple of hundred years of carefully selected breeding by generations of horsemasters, who have been without peer in the world for judgment, knowledge and ability, and who have moreover been prepared and able to lavish uncounted fortunes on the pursuit of their aims, have produced a horse as nearly perfect as it is possible to achieve.

Apart from the quality of speed, in which the Thoroughbred far surpasses any other breed, the best specimens of the race possess all the other desirable qualities that should be present in a riding horse of the highest class. First and foremost, the Thoroughbred's action is the most desirable from a riding point of view; his walk is long, free, easy and fast; his trot roomy, rhythmic and balanced; his gallop the finest pace of all, with a long, sweeping and incredibly easy stride.

Whereas his great speed would not be possible without this perfect action, the action in turn would not be possible without equally perfect conformation. Conformation is, I think, a most difficult thing to convey in words, and in a sense it is an ungrateful task to try and do so, because such descriptions are at once superfluous for those who have an eye for a

39

horse, and who can spot the points when they see them, and unintelligible for those who lack that ability. To be able to see, at a glance, the telling points of a horse which together make up his conformation, good, bad or indifferent, requires much practice and the development, almost, of a special sense. It seems to me that too much detailing of the points of a horse is confusing rather than helpful to the student, and instead therefore of offering the customary catalogue of points and faults, I prefer to rely rather more on conveying the image of perfection by means of pictures of some really high-class animals.

But without going into any great detail, I would yet like to try to convey a feeling of the qualities that are essential in the make-up of any great horse.

And the first of those qualities and the most essential of all is *Quality*, pure and simple. Quality is the impression one obtains at first sight, in the presence of the real equine aristocrat; it is in the manner of his approach, in the pride of his bearing, in the outlook of his bold and confident eye; it is in his finely cut, expressive head, in his alert and shapely ears, in the play of his muscles under a fine and silky skin; in short, it is in the hallmark of breeding, that will be written all over him.

And the second of those qualities is *Balance*, which is true symmetry of proportions; it is the quality that enables the horse to be light and graceful, as he stands and as he moves, always airily, and without apparent strain or effort.

In effect, quality and balance merge into one another, and when combined to a high degree in the same animal, there can be little doubt but that such is a great horse, and that he will be able to display that supreme fluency of movement that distinguishes the Thoroughbred above all other breeds. For the Thoroughbred's action, though full of energy and brilliance, and beautiful to behold, is markedly low, and with little knee-action; he is able to bring his toe forward further than any other horse, and to make a longer stride at the expense of a lesser effort. Whereas the common-bred horse can only gallop much more laboriously with pronounced knee-action, a much shorter stride and consequent loss of speed, the Thoroughbred is able to do so much better because of his superior front. Whereas the common horse moves mainly from the elbow, the Thoroughbred moves from the shoulder.

It is the shoulder, the upper arm, the chest and the head and neck, which together form a horse's front. The shoulder should be fine, long and sloping, and be marked with a pronounced wither; in my opinion and experience the angle of the upper arm is of the utmost importance, as it can make or mar the freedom of a horse's action, even though the shoulder itself might appear good. The upper arm lies inside the horse's body, and is the connection between the shoulder and the forearm. If this upper arm is situated too horizontally, the forearm will appear to emerge from the body rather far behind the chest, and in that position the shoulder's free action will be found to be impeded. I think it is essential that the upper arm should be rather steep, so that but little chest appears in front of the forearm; the shoulder will then have that beautiful freedom of movement that is both so desirable and so essential. The chest should not be mean and narrow, as there will then be no proper room for heart and lungs; a narrow-chested horse is but a weed. On the other hand we do not want a great wide carty chest, which is a sign of common breeding and makes the horse heavy. The head should appear, and be, light to carry, and the neck fine and long. The head and neck play a most important part in the balance of a moving horse, and a short-necked horse is at a distinct disadvantage in this respect.

To be a good mover is all-important; movement begins from the shoulder, and the shoulder should be seen to move freely; the good mover will throw his toe out markedly, and with some cadence, at the walk, and the trot and the canter, and do it all with little action of the knee; exceptional movers will sometimes throw the toe forward with such energy that one can actually see the sole of the foot. Horses that cannot use the shoulder and that move from the elbow are called elbow-tied; they are quite unable to produce the

41

long and energetic stride that has been described, and they will accordingly be an inferior ride and lack speed, yet that mechanism alone would not be much good without equally first-class motive power, that is provided by a stout heart and a first-class pair of lungs. To house these organs the horse requires a body of great depth, with well-sprung ribs, and a great girth. The best type of mature Thoroughbred will look well let down, and his girth may measure as much as 70 inches or even more.

To round off the picture, the Thoroughbred possesses the indomitable courage needed to face and to sustain great exertion; it is when things get really difficult and hard that his persevering will to fight on comes to light. However long and tiring the effort, he will maintain his qualities of gaiety and go until the end, in fact, if necessary, until he drops.

That is what makes him, apart from being the greatest racehorse, such an exhilarating ride for any rider whose skill is of the right temper and a match for his exuberant courage.

To be recognized as a Thoroughbred, a horse must have been registered in the General Stud Book. To be eligible for registration a horse must be *either* able to be traced down all lines of its pedigree to horses already registered before 1st January 1980 in a) The General Stud Book, b) The Stud Books of other named major thoroughbred breeding areas of the world *or* to prove satisfactorily 8 recorded crosses consecutively with horses qualified as above, including the cross of which it is the progeny. Animals up-graded from the non-Thoroughbred register are placed in the Appendix to the General Stud Book. These animals normally need to show performances in races open to Thoroughbreds, to warrant their assimilation.

The first edition of the General Stud Book was published by Mr. Weatherby in 1791. But that must not be taken to mean that the breed itself, or at least the foundation of it, did not exist before that date. Efforts at improving the native breed by the importation of stallions and mares of Oriental and Spanish origin can be traced back well into history, but the information available about these earlier importations is insufficient to serve as a basis for any reliable theory.

42

It is known, however, that James I maintained a stud of racing horses, and that he had there some half dozen Arabians, imported by Sir Thomas Esmond, one of whom was known as the Markham Arabian. Cromwell too was interested in breeding, and maintained his own stud with, amongst others, the Oriental stallion, White Turk.

It was Charles II, however, who appears to have given the real impetus to the development of what was to become the modern Thoroughbred. He sent his Master of the Horse, Sir John Fenwick, on a mission to the Levant for the purchase of some of the best Oriental mares he could find. Sir John came back from Tangier with the so-called 'Royal Mares', Barb, Arabian and Turk; he brought some Oriental stallions also.

Subsequently we find steady importations of Oriental blood, and Volume II of the General Stud Book contains the names and pedigrees of more than 200 Oriental stallions and mares, for the period 1721-59.

Although some 176 of these were stallions, there are three that are particularly noteworthy for their permanent influence on all subsequent Thoroughbred stock. They were the Byerley Turk, a contemporary of William of Orange; the Darley Arabian, imported during the reign of Queen Anne; and the Godolphin Arabian, who died in 1753 at the age of 39 years.

It is notable that all Thoroughbred stock trace back to these three sires in the male line: the Byerley Turk was responsible for the Herod line, the Darley Arabian for the line of Eclipse, and the Godolphin Arabian for the Matchem line.

In addition to these three, there are other Oriental stallions who, though not in the direct male top line, have exerted great influence on Thoroughbred stock; the most notable of these are the Leedes Arabian, whose name appears in a larger number of pedigrees than that of any other stallion, and the Alcock Arabian, who appears in the pedigree of every grey Thoroughbred now in existence, and whose colour has persisted without a break for more than 200 years.

Bruce Lowe, whose breeding theories are well known, has classified the entire Thoroughbred race into 43 families, according to the tap-root mares wherefrom they are descended. According to Lady Wentworth, the number of distinct families is smaller still, as she has found some of the Bruce Lowe families to have merged into one another; in her opinion, there are only 25 families, and of that number only 18 that are of real importance. It is at any rate quite certain that the number of tap-root mares is no more than, in the words of Lady Wentworth, a mere handful.

What is equally certain is that the Thoroughbred of today is predominantly Arabian in descent. Lady Wentworth, who has made a close and deep study of this subject, publishes figures representing the number of repeat crosses of foundation sires that she has found in the pedigrees of some of the recent winners of classic races; in her tables she also includes the Arabian foundation mare Old Bald Peg, whose influence on pedigrees appears to have been more pervading still than that of any of the sires. In compiling her tables, she has calculated back over 26 generations, and anyone familiar with the story of the chessboard and the grains of wheat will understand the astronomical figures arrived at.

However, her table is so illuminating of the preponderant influence of Arabian blood in the modern Thoroughbred, that I cannot do better than reproduce it on the opposite page.

The Darley Arabian is at the head of the direct sire line of both Sun Chariot and Big Game, whilst the Arabian mare, Old Bald Peg, is Big Game's tap-root mare; it is remarkable that more than 300,000 repeat crosses of her blood appear in his pedigree.

The greatest credit is due to English breeders who through unrivalled skill and splendid management, have created this supreme breed of horses out of the beginnings sketched above. No praise can be too high for their achievement.

But let us not forget that the excellence of the Thoroughbred could not have been attained without the free admixture of the one blood that is, amongst all others, the fountain-head of all improvement in horse-breeding, that of the Arab horse.

For it is not open to doubt that it is to him that all warm-

FIGURE 5. LADY WENTWORTH'S TABLE OF REPEAT CROSSES OF ARAB BLOOD IN THE PEDIGREE OF CLASSIC WINNERS.

	GODOLPHIN ARABIAN	DARLEY ARABIAN	BYERLEY TURK	LEEDES ARABIAN	DARCY WHITE TURK	DARCY YELLOW TURK	LISTER TURK	HELMSLEY TURK	OLD BALD PEG ARABIAN MARE	SULTAN ARABIAN
BIG GAMB	28,232	44,079	64,032	187,197	90,420	294,508	63,832	109,946	367,162	367,162
SUN CHARIOT	40,279	68,394	94,706	169,899	230,307	279,293	60,717	102,912	233,579	233,579
HYPERION	15,594	23,998	32,199	31,426	118,182	111,938	22,350	41,003	138,827	138,827
FAIRWAY	14,931	22,734	33,188	58,350	50,252	92,804	18,662	34,478	113,740	113,740
WINDSOR LAD	25,376	38,944	61,783	93,573	103,753	152,544	30,366	56,880	184,091	184,091

(From: *The Authentic Arabian Horse*)

blooded breeds of the world owe the best of their blood. The warm-blooded horses originate from the Near Orient and the countries around the Mediterannean basin, as against the cold-blooded horses, who are natives of North Western Europe and parts of Northern Asia. The warm-blooded breeds are characterized by lightness, quality, fineness, courage and impetuosity, the cold-blooded breeds by massive size, heavy bone, big heads, a certain amount of coarseness and phlegmatic disposition. Where most of the so-called light horses belong to the warm-blooded variety, the big agricultural and carthorse breeds are cold-blooded.

Throughout the Middle Ages the Mediterranean peoples, Arabs, Turks and Moors had the best light horses. During the centuries of Moorish dominion over Sicily, Southern Italy and Spain, they took their excellent horses with them and created or further improved the Oriental type of warm-blooded horse along the northern shores of the Mediterranean Sea. When the Moorish Empire receded to give way to the rising might of the Spanish Empire, it is certain that they left a most excellent breed of horses behind them. And all through the period of Spain's world power, and much later still, and well into the eighteenth century, the Spanish, Andalusian and Neapolitan horses were famed the world over, and were in great demand at every Court in Europe. It is certain that these horses, themselves replete with Arabian blood, have left their mark and influence on the light-horse breeds of every country.

All these horses were preponderantly Oriental in character, with a strong predominance of Arabian blood. But then the use of Arab blood for the improvement of native breeds has been continued right up to date in all countries where importance is attached to the improvement of native breeds. Russia, Germany, Poland, Austria, Hungary, Yugoslavia and France all maintain considerable studs of pure-bred Arab horses. Many of these studs are either State owned or State subsidized, and stallions are sent out during the breeding season for the express purpose of covering native mares. This extensive use of the Arab is based upon the recognition that no other breed or type of horse has the same astonishing power of transmitting its characteristic qualities with such amazing certainty to its offspring. The Arab is proverbially sound in wind, eyes and limb, and experience proves that in many cases one single infusion of Arab blood suffices to outbreed all sorts of native defects and unsoundness. Moreover, the Arab is sober and exceedingly hardy, and no horse in the world, not even his speedy descendant the English Thoroughbred, can equal his endurance and staying-power. With the exception only of the Thoroughbred, who has been bred specially for speed, he is the fastest horse on earth. Although of comparatively small stature, he is amazingly strong and powerful, and able to carry a lot of weight. In quality, he is of course supreme — beautifully proportioned, marvellously balanced, and exceedingly good to look at. His temperament, though hot and fiery, is of great gentleness, and his intelligence is much superior to that of any other breed.

The origin of the Arab is lost in antiquity and in fable. The Arabs recognize five great strains of blood, which are together called the Khamsa; according to some accounts, these strains descend from the five mares of Solomon, and others have it that five mares were selected by the prophet Mohammed and marked by him with his thumbmark.

It is certainly true that Mohammed recognized the immense value of a first-class breed of horses for the life of his peoples in the vast wastes of the Arabian Desert, and the Koran abounds with references to the horse, and with commands in the matter of the care and affection to be bestowed upon him.

The so-called Prophet's Thumbmark appears in a slight irregularity in the growth of the hair, usually on or near the crest of a horse's neck; the mark has the appearance of a small circle, with the hair radiating out from the centre, which looks as if it had been flattened out by an impression, and the term 'thumbmark' is certainly an apt description of it. This mark occurs on certain horses only, and is fairly rare. The Arabs believe that a horse so marked is a particularly good one; whatever foundation there may be for that belief, I can only say that the horses that I have had, and which did possess this distinguishing mark, were in fact extremely good and genuine animals. Alternatively some people maintain the Prophet's thumbmark is a thumb sized indentation usually found at the base of the neck.

Apart from the five great strains of the Khamsa, which are the most esteemed as the purest of the pure, there are sixteen other strains of pure blood which are valued almost as highly, and most of whom have a great deal of Khamsa blood in their veins. However, the Arab takes his pedigree, or strain, from the mare, and if a mare of one strain be mated with a stallion from another, the produce will belong to the mare's strain.

The five strains of the Khamsa, each of which has a number of sub-strains, are the following:

1. The 'KEHILAN' (feminine 'Kehileh' or 'Kehilet').

These are the most numerous breed in Arabia, and very highly valued; they are considered the fastest breed, and are frequently somewhat bigger than other strains, though not as a rule the most beautiful. In type they are the most nearly resembling the English Thoroughbred, which may be due to the influence of the Darley Arabian, who was a Kehilan. The Kehilan is thought to have influenced the other four strains of the Khamsa a great deal.

The word 'Kehilan' appears to be connected with the black colour of the skin, that is the hallmark of a pure-bred Arabian, whose skin shall be black no matter what the colour of his coat; gradually the meaning of the word has come to be synonymous with our expression 'Thoroughbred'.

The most favoured sub-strains are the Kehilan Ajuz, Kehilan Nowag, Kehilan Abu Argub, Kehilan Abu Jenub and Kehilan Ras-el-Fedawi; other well-known sub-strains are the Dajani, Rodan, Jellabi, Momrahi, Kahtan, Al Wadaj, and some of these sub-strains are subdivided again.

2. The 'SEGLAWI' (feminine 'Seglawieh').

There are four sub-strains of this blood, all identical in

origin, since descended from four sisters, the Jedran, Obeyran, Arjebi and el-Abd. The Seglawi Jedran is considered the best of all Arab horses, but the strain is very rare and it is held the only strain of absolutely pure Seglawi blood, the other three having been crossed with Kehilan.

3. The 'ABEYAN' (feminine 'Abeyeh').
This strain is small in stature but very handsome. The most favoured sub-strain is the Abeyan Sherrak, of which only a very few pure families now subsist.

4. The 'HAMDANI' (feminine 'Hamdanieh').
The Hamdani are another uncommon breed, but one with great quality and a tendency to producing greys. There is only one sub-strain that is recognized as pure, namely the Hamdani Simri.

5. The 'HADBAN' (feminine 'Hadbeh').
The Hadban completes the five breeds of the Khamsa; it is not a numerous strain, and was mostly in the possession of the Roala tribe only. There are three sub-strains: the Enzekhi, which is the most prized, the Mshetib and the El Furrd.

Some of the better-known strains, that do not belong to the Khamsa, are the Jilfa Stam, Managhi, Dahman om Aamr, Saadan Togan, Greban.

Pure-bred Arabs have a black skin and are always true coloured, and two-coloured horses such as skewbalds and piebalds do not occur. Black horses are rare, and very highly prized; the most predominant colour is bay, and the belief is that bay horses are hardier than any; greys are usually born black, or nearly so, and become gradually lighter in colour and may end up by being pure white; chestnuts are supposed to be the speediest.

Apart from the European countries that have already been mentioned, several of the South American countries are regularly purchasing Arabians for the purpose of improving their native breeds. In the U.S.A., too, there are now a number of well-founded Arabian studs. Since the Arabian is undoubtedly the fountain-head of the best blood in the world, and has no equal in the power to improve other breeds, it is obvious that there will be a worldwide demand for Arab horses of the right type as long as light-horse breeding continues.

From that point of view, the English studs of Arabian horses, whereof there are quite a number and of whom some have bred consistently stock of the highest class, are of great importance. Also, in England itself, Arabian blood is in increasing demand for the purposes of breeding Anglo-Arabian and part-bred Arabian horses, following possibly the example of France, where the Anglo-Arab has become one of the leading breeds and the one that is more than any other in demand for high-class riding horses.

Prior to 1921, pure-bred Arabians, whether imported or bred in this country, were eligible for entry in Messrs. Weatherby's General Stud Book; since that date, however, the keepers of the studbook have decided that no further Arabian horses will be accepted for entry unless they can be traced back to a strain already accepted in the earlier volumes of the Book. Unfortunately this too was discontinued as from Vol. 36 of the General Stud Book.

The Arab Horse Society, established in 1918, is at present the recognized authority in this country, and registration in the A.H.S.B. warrants any horse so entered to be a pure-bred Arabian.

Anglo-Arab is the designation of any Arab-Thoroughbred cross, and it does not matter whether the sire is Arab and the dam Thoroughbred, or vice versa; but it is most generally held, and I concur with that opinion, that the Arab stallion on a Thoroughbred mare is likely to yield rather better results than the covering of an Arab mare by a Thoroughbred stallion. Such produce is of course eligible for registration in the Anglo-Arab register of the Arab Horse Society.

In France the breeding of Anglo-Arabs is carried out on a very considerable scale, and the progeny are in great demand. It is recognized there that the Anglo-Arab is almost

50

invariably an exceptionally fine jumper, and the bulk of France's famous international show jumpers have been Anglo-Arab. The Cavalry School at Saumur has also been partial to the breed, and many of the high-school horses of the Cadre-Noir belong to it.

The cross has a great deal to recommend it: from the Thoroughbred it inherits size, range and speed, and from the Arab much of his incomparable bearing, endurance, and above all, intelligence and temperament. The Anglo-Arab is definitely easier to school than the Thoroughbred because of his superior intelligence, but particularly because of his better temperament. The one fault to find with the Thoroughbred is his generally very highly strung temperament, which causes him to be nervous, excitable and easily upset. The Arab, though at least equally fiery and hot-blooded, is much more levelheaded; he takes things more calmly and does not rattle himself to anything like the same extent. I have often thought that this extreme nervousness of the Thoroughbred may be due to the accumulated strain of racing, for generations, at much too young an age. However, it is certainly true that the average Anglo-Arab is more amenable than the average Thoroughbred, and for that reason often makes better hunters, hacks or dressage horses.

Part-bred Arabs, which may be registered as such in the Part-bred Arab Register of the Arab Horse Society, are horses with an authentic strain of Arab blood. Usually they are the produce of an Arab stallion out of a mare other than Thoroughbred; consequently there is likely to be a great variety in size, type and shape in this particular kind of horse, since an Arab stallion may be mated equally well with a pony mare of 13 hands 2 inches as with a hunter of 16 hands 2 inches. And the produce of both matings may be equally attractive and suitable for its particular purpose. At my stud I have bred, from the same sire, children's ponies of under 14 hands 2 inches and great big hunters standing 16 hands 3 inches, and up to 14½ or 15 stone.

The most striking experience that I have had with these crosses is the exceptional degree of quality achieved in the produce even of indifferent and sometimes downright ugly mares; and in every case the sire has given them that wonderful head carriage that is always just right and which, in the produce of a good Arab, requires no further making.

The English Thoroughbred, the Arab, and the Anglo-Arab are the only breeds in Britain to whom the designation 'thoroughbred' can be applied justifiably, on account of their ancient and flawless ancestry.

There are one or two other breeds, such as the Hackney and the Cleveland Bay, to whom the designation 'pure' or 'pure-bred' might be applied with equal justification, because the studbooks of these breeds, although of much more recent date, trace back to recognized foundation stock, and limit the introduction of any blood foreign to the breed.

Apart from that there are other studbooks, such as the Non-Thoroughbred Register and Hunter Stud Book and the National Pony Stud Book, that admit for registration individuals considered to be of the desirable type; frequently the ancestry of such individuals is either unknown or unauthenticated, as for instance when an individual is stated to be by such and such a stallion out of a dam by some other stallion. Although studbooks of this kind are of great value in encouraging the breeding of animals of an approved type, so that the fact of horses being registered therein is not without importance or value, the stock entered in them can yet not be considered to constitute a breed, and the correct designation to give to such stock would be to say that they are registered in such and such a studbook.

Having made this clear, we can now discuss the Hunter.

The term 'hunter' is somewhat difficult to define; for though there is a Hunter Stud Book, there is yet no such thing as a hunter breed. The term 'hunter' is used frequently in a functional sense as applying to a horse used for hunting; in that sense the expression 'hunter' may cover a large variety of horses, from a high-class thoroughbred down to a moor-land pony, from the smart blood-horse, fit to shine in a fast hunt behind one of the crack packs in Leicestershire, to the ride-and-drive cob that may be quite good enough to potter behind hounds in some rough provincial country.

But the true significance that a breeder, or a judge, would attach to the term 'hunter' is one of type. In their interpretation, the hunter is the type of horse likely to be able to follow hounds over a big, fast country, a horse that can carry weight and one that will stay; one, therefore, that possesses both speed and power. Such a horse will be of the sort that hunting people would recognize as a Leicestershire Horse, meaning thereby that he would be fit to be hunted in that country. To be able to do so, the horse requires much blood, as without it he would be able neither to follow nor to stay the pace; combined with blood goes quality, and combined with quality go looks.

Though the hunter needs to be fast, he does not necessarily require the speed of a racehorse; actually, the hunter will usually gallop well within himself, and that is very necessary since otherwise he would never be able to go the considerable distances that are often required. But he has to carry, as a rule, considerably more weight than the racehorse, and with that to gallop and to jump on going that may be far from ideal, and his days may be very long. He must therefore be robust, with a strong back, the best of loins, a good deep frame well ribbed up, and stand on the best of limbs, with plenty of bone and good open feet, He also must have size, though preferably not outsize.

Hunters are subdivided into lightweight hunters, to carry under 12 st.; middleweight hunters, to carry up to 14 st.; and heavyweight hunters, to carry over 14 st. About 16 hands 2 inches is probably the most desirable size, although this will depend on the height of the rider. Anything from 16 hands to 17 hands is quite admissible. Horses over 17 hands tend to become rather outsize, unless well made. Horses under 16 hands are hardly of the true hunter type, excepting perhaps as ladies' hunters up to 12 st., who may measure as little as 15 hands 3 inches.

It goes without saying that the best type of Thoroughbred horse, one with plenty of substance and size, makes an ideal hunter type, and provided their temperament is right and not too nervy, they are as a rule hard to beat as lightweight hunters. Thoroughbreds that would be considered up to more than 13 st. are rare, and Thoroughbreds up to more than 14 st. exceedingly rare indeed, and exceedingly valuable.

But a hunter need not be Thoroughbred, and as a rule will not be, although the best horses are usually very close to the blood, by a Thoroughbred sire out of a hunter mare, herself by a Thoroughbred sire, and so on. Since lightweight hunters are usually closer to the Thoroughbred, if they are not actually in the book, they are easier to breed with plenty of quality; and easier to come by than the heavier horses. It is much more difficult to produce a quality middleweight horse, and very difficult indeed to succeed with a weight carrier. Usually the introduction of more common blood is resorted to, at the expense, of course, of quality. I shall say more about this in the chapters on breeding.

Any hunter of the best type, whether lightweight, middleweight or heavyweight, must have the Thoroughbred's true type of riding action, dependent, as we have already seen, on shape and freedom of the shoulder; the action must be free, easy and low, and the stride long; short action and bent knees, particularly at the gallop, are a sure sign of common ancestry and unpardonable in a hunter.

There is not much in the matter of colours, provided they be true colours and not washy; too much white would not be favoured in a hunter, and two colours, such as skewbalds and piebalds, are not in accordance with first-class breeding.

Any hunter that answers the requirements set out in the preceding paragraphs is likely to be a good horse and almost certain to be nice to look at, although not necessarily fit to show. There are a great many very nice and even beautiful horses that would be a credit to anyone in the most select hunting field, yet that might fail to win in the show-ring because of some imperfection of conformation. Showing is largely a matter of comparing horses, and naturally the one that comes closest to the ideal that the judge is bound to have in his mind is going to be the winner.

Judges vary a good deal in their ideas, and some may attach a great deal of importance to details that others might be prepared to overlook. Consequently, no hard and fast rules

54

can be laid down. I think, however, that the photographs of prizewinning hunters of recognized merit may go some way in clarifying the desirable points of a hunter; I would like to add my opinion that action and quality are far more important than detail of conformation, since a horse that cannot move is useless as a hunter, and one without quality liable to fail in genuine courage.

The Hack also is a type of horse, and not a breed, since he might, in theory at least, be of any breed; there is no such thing as a Hack Stud Book.

The hack is the sort of horse most suitable to be ridden for the mere pleasure of riding, pleasantly, quietly, effortlessly and elegantly, in country or park. It is generally conceded that for that kind of riding a somewhat smaller, lighter, and finer type of horse is rather more pleasant and less strenuous than the stronger and bigger hunter. Since hacking is normally restricted to a mild exercise of an hour or two's duration, the horse does not require anything like the same substance and strength as is expected of the hunter.

But to deserve the description of a hack, in the sense in which the word is used to describe the class for this type of horse at our leading shows, the horse must be full of quality, have a small and almost perfect head, well set on and nicely carried on a well-curved neck that should be slender, and that must on no account be thick and coarse; the animal must be an outstanding mover, attractive and even pretty. He may be somewhat flashy, and a good deal of white, which would not be appreciated in a hunter, will be no detriment to a hack. Anything at all common or coarse will at once spoil a hack. It is almost essential therefore that he should be a Thoroughbred, or very nearly so, or Anglo-Arab, as otherwise the quite essential degree of quality could not be obtained.

But he must be of the smaller, finer and less powerful thoroughbred type, of a height of from about 15 hands to 15 hands 3 inches; yet though fine, he must have some substance and a good middle, and he must on no account be what is commonly called a 'thoroughbred weed'.

The usual classes recognized at our leading shows are for

hacks exceeding 14.2 h.h. but under 15 hands, and for hacks under 15 hands 3 inches. The bigger type of hack and the lady's lightweight hunter are apt to merge into one another as a type, and some horses appear almost equally suited to either class. But a real hack should not be shown in a regular hunter class, where he will appear too small and insignificant in comparison with the normal hunters of about 16 hands 2 inches.

Neither should the real hack, of the 15 hands 3 inches variety, look the hunter type, for if he does he will be too coarse. It is of equal importance that the smaller hack, of 15 hands or under, although of pony size or almost so, should not look a pony, and on no account must he move like one and have pony paces. Even the small hack should move and carry himself with the full dignity of a horse.

The term 'Cob' is merely a name, indicating a stockily-built, thickset, small horse of pony size, that is not over 15.1 h.h. Classes for cobs are normally divided into Lightweight up to 14 st. and Heavyweight over 14 st. The typical riding cob is in reality nothing more or less than the equivalent of the ride-and-drive horse used by farmers in olden days to go to market with. The contemporary show variety are supposed to be ideal mounts for heavy old gentlemen to follow hounds on, but in reality they appear to be more of a show-curiosity than anything else. It is true that some animals of this type are quite attractive in their own way, but there is too much of the harness horse in them to make them pass muster as a true riding horse; their action is against them, since they bend the knee too much. The quality of the show cob owes so much to expert grooming and trimming that without such attention the difference between such an animal and the better type of vanner might not always be very apparent. I do not pretend to know how a weight-carrying riding cob is or can be bred, and I have never been fortunate enough to meet any breeder able to enlighten me on the matter, although it may be that there are some who do know. However, failing convincing proof to the contrary, I shall continue to assume that show cobs are an accident, and may be either a hunter gone wrong or else a vanner turned out above expectations. It is the fashion that cobs shall be hogged. No great exception can be taken to hogging, since it is a way of improving the looks of a common, short and heavy neck.

In the Hackney's case there is no need to resort to hogging, for his neck has all the quality and fineness of a truly well-bred horse. He can claim to constitute a true and pure breed, with an ancestry tracing back nearly as far as that of the Thoroughbred.

It was towards the end of the nineteenth century that the hackney, found recognition as a breed, and was brought to prominence by the Hackney Horse Society who held their first show in 1885. There were 123 entries for that show, but the popularity of the breed was so much on the increase that 383 entries were received for the event in 1893 for the event in 1893 and 442 for that of 1896.

The hackney traces his descent undoubtedly from the original type of excellent roadsters that used to be bred in the Eastern Counties and in Yorkshire; we may safely assume that these trotting roadsters received a liberal admixture of Spanish blood from the quality horses which were, as we know, imported from time to time in considerable numbers.

The undoubted quality and true blood characteristics of the breed are due to the influence of Arabian blood upon an already good foundation, and the hackney shares with the Thoroughbred the illustrious ancestry of the Darley Arabian. The foundation stallion of the present studbook is Old Shales, the sire of Scots Shales, who stood at a fee of one guinea and a shilling the groom. He himself was a grandson of Blaze, who was foaled in 1733 and was the son of Flying Childers, out of Grey Grantham by the Brownlow Turk, the latter's dam being by the Duke of Rutland's Black Barb.

Of the many sons of Shales, Driver and Scots Shales became the pillars of the studbook, and it is from them that most of today's best horses are descended.

From the early days the hackney has been a trotting horse, and a very fast trotter at that; it is recorded that Old Shales could trot 17 miles in the hour; a daughter of Driver covered 15 miles in the hour, carrying 15 st.; Read's Fire covered a mile in 2 minutes 49 seconds; in the year 1800 the mare Phenomena, at the age of 12 years, covered 17 miles in 53 minutes. She did 4 miles in 11 minutes and, when 23 years old, was still able to travel 9 miles in 28½ minutes. This famous mare was by Othello, and at one time in her career became the property of the Duke of Leeds for the sum of 1,800 guineas. She was only 14 hands 2 inches high.

That the modern show hackney has lost none of this capacity for fast travel may be illustrated by my own experience during the years of war with Holywell Squire. This horse, who is by the outstanding 'modern' sire Bertrano, was at one time a famous performer in the show-ring, the only horse on record to have won the big harness horse class at Olympia three years running, and many other trophies besides. Though then 15 years old, this horse still won some of the most coveted trophies at wartime shows. But his main war work was on the road, where I used him a great deal in order to save petrol. On long journeys, driven easily on a loose rein, and allowed to choose his own pace, jogging along quietly with plenty of walking, he never fails to average 12 miles an hour and never turns a hair. I have an outlying farm at exactly 8 miles from my place, and Squire does that journey a couple of times a week, in a gig or a four-wheeled wagon, in exactly 30 minutes; and he does that without any urging, well within himself, and including time lost in walking up and down a few short hills.

But then the trot is the hackney's best pace, and no other breed of horses can equal that trot for perfection of rhythm and balance; the hackney's trot is at the same time powerful and yet extremely light and graceful; he appears hardly to touch the ground, his hoofbeat is light and airy, and he floats, as it were, from one diagonal to the other, with a moment of suspension distinctly marked. His balance is supreme, hind feet treading far forward and well underneath the body, head high, bent at the poll, neck arched gracefully and so light in front that no weight at all appears to limit the spectacular brilliance of his movement in front. The good hackney moves not only very high, but also very freely with a tremendously extended stride. His movement behind is almost as brilliant as that in front, with hock action of unequalled energy. The front action is entirely from the shoulder, which is long, fine and muscular, and free from fat, in fact a shoulder as good as that of the best type of thoroughbred. The hackney's body should be deep, powerful and short, with strong loins and powerful quarters, legs rather short, and the horse, though short in the back, should yet stand over a lot of ground. The neck is long, fine and slender, carried high and beautifully curved; the head, though rather larger than that of the Thoroughbred, is yet full of quality, lean, well chiselled and without coarseness, with a broad forehead and big confident eyes, set well apart.

The most striking characteristic of the hackney, when not in movement, is the inimitable way in which he stands on the ground, firmly and boldly, squarely on all four feet and looking as if he meant to anchor himself there, as solid and as sure as a statue. He is a true blood-horse, with unbounded courage, full of fire and generous in the extreme; it is a delight and even a thrill to drive him.

The most usual height of the full-size hackney horse is from 15 hands to 15 hands 3 inches, although occasional animals measure up to 16 hands 2 inches. In addition we have the hackney pony, of 14 hands or under. Favourite colours are chestnut, brown, bay and black; four white feet are valued, and are often met with in the best animals.

The breed had its heyday from about 1885 to about 1914, but since the latter date the harness horse has been eliminated gradually, and in the end almost totally, by the motor car. Though in 1939 some stalwarts still stood by the breed and enlivened the big shows with their exhibits, it is undeniable that they were few in number, and accordingly the breed itself was sadly reduced in numbers too. But with the increase in the popularity of driving more people are breeding and showing hackneys. So this magnificent type of horse is unlikely to disappear altogether. Since the war years there has been a considerable revival in the interest taken in the driving-horse and a wider appreciation of the pleasure that may be derived from him, and the hope that this may continue and so bring the hackney breed to new vigour is perhaps not entirely unjustified.

The Cleveland Bay is another recognized breed of harness horse; their action, though very good, misses the height and brilliance of that of the hackney; the breed enjoys popularity as a high-class carriage- and coach-horse, for use in the rather heavier type of carriage. The Cleveland is a good-looking horse, built much on the lines of the Thoroughbred, but of course rather heavier, taller, and not quite so blood-like. The breed hails from Yorkshire, which was predominantly a grass county, where no heavy plough-horse was required. For some two hundred years or more they had possessed in that county a very useful type of strong and active, medium-heavy horse. Upon this strain of horses Thoroughbred and Arab blood were introduced about the beginning of the eighteenth century, commencing apparently with Manica, a half-bred Arab by the Darley Arabian, and with the Thoroughbred horse Traveler, who was foaled in 1753. Another well-known ancestor is Jalop, who was by a son of the Godolphin Arabian.

The breed went through several periods of ups and downs, the downs being accounted for by the introduction of cold blood during periods when the country turned to corn-growing, or when for other reasons heavy draught-horses were particularly in demand.

Towards the end of the eighteenth century America began to demand Cleveland horses, and from that moment a definite revival set in, marked by the foundation of the Cleveland Horse Society in 1884.

The Cleveland's colour should be bay, preferably with a golden shine, with black points and black legs. He is a horse of excellent conformation and of strong constitution, a real utility horse that will do well on the land and yet one that will disgrace not even the finest gala turnout.

To be considered a pony, a horse must not exceed 15 hands

By 'Basa': 'Alba', three-year-old Anglo-Arab filly. First and reserve champion, Arab Horse Show 1947.

Modern three-year-old Anglo-Arab gelding 'Redwood Replica'.

Heavyweight hunter 'Darrington'.

Modern heavyweight hunter 'Flashman'.

Middleweight hunter 'Beau Geste'.

Modern middleweight hunter 'Dual Gold'.

Champion show hack 'Liberty Light'.

Modern Champion show hack 'Rye Tangle'.

Modern show cob 'Buzby'.

Champion Hackney 1938 'Barcroft Belle'.

Modern Cleveland Bay 'Masterful Jack'.

Polo pony 'Valentine'.

Modern riding pony youngstock Champion 'Bradmore Nutkin'.

Modern working hunter pony 'Towy Valley Moussec'.

Modern Welsh mountain pony stallion 'Coed Coch Rhion'.

Modern New Forest pony stallion 'Deeracres Franco'.

Modern Exmoor pony stallion 'Dunkery Buzzard'.

Modern Dartmoor pony stallion 'Allandale Vampire'.

Modern Shetland pony Champion 'Bincome Venture'.

Modern Highland stallion 'Dallas of Stanford'.

Modern Dales mare 'Brymor Mimi'.

Modern Fell mare 'Bewcastle Bonny'.

Champion Suffolk stallion Morston Gold King'.

Champion Shire stallion 'Silwood Streamline'.

Champion Clydesdale stallion 'Hawkrigg Headline'.

Champion Percheron stallion 'Berkswell Grey Victor'.

A striking produce group. 'Mrs Bee', the white mare on the right, with seven of her eight offspring, one, two, three, four, five, six and seven years old. A mare, not perhaps of the show type, but roomy, well ribbed up, good feet, rather chesty. But the head is outstanding, forehead tall above the eyes, eyes large and very far apart, fine muzzle. She has transmitted this feature, and excellent limbs and feet, to all her offspring, who together may be the most striking group of young hunters, out of the same mare by different sires, ever seen.

Pony mare and foal: part-bred Arab mare 'Dusky Lady' by 'Ruskov' out of a Welsh pony mare, with filly foal by 'Basa'.
(Property of Miss de Beaumont)

Premium stallion: 'Jean's Dream', a splendid type of thoroughbred hunter sire.

Modern premium stallion: 'Senang Hati'.

in height, and any horse under that size, with the exception of the strong thickset variety that are known as cobs, is a pony. Childrens' ponies are usually limited to 14 hands 2 inches except for side-saddle and Working Hunter ponies when the maximum height must not exceed 15 hands. Classes are included for younger children on smaller ponies.

The National Pony Society publish a register of Riding Ponies which must not exceed 15 hands at maturity. This society was formed from the old Polo Pony Stud Book Society.

The best type of polo pony is a small horse with much quality and blood, and with considerable substance, as he must be able to carry weight; he must also be fast and very well balanced in order to be suitable for this fast strenuous and exacting game, with its continuous stopping, starting and turning. The Polo Pony Stud Book Society was formed at the end of the last century in an attempt to breed the ideal animal for the game. The Society accepted suitable horses or mares provided they answered the requirements for size and type, and provided they were from registered parents or by a Thoroughbred, or Arab sire out of a registered dam.

In my opinion balance is the most essential characteristic of a first-class polo pony; that means a short-backed horse, with strong loins, good quarters and a straight hindleg, prominent withers, a good shoulder and really good head carriage; a horse so built will be light in front and thus better fitted for his special work. On account of size type, balance and headcarriage, the Arab is eminently suitable as a polo pony sire. However in recent years small Thoroughbred horses are favoured. Polo ponies never have constituted a true breed. British ponies of the Mountain and Moorland breeds, on the other hand have their own stud book societies, and since 1985, their own stallion licencing schemes.

The principal types or breeds of Mountain and Moorland ponies, the native breeds of these islands, comprise the Welsh, New Forest, Highland, Fell, Exmoor, Dartmoor, Dales and Shetland ponies.

It is difficult or impossible to trace the correct origin of all the various types but it may be assumed safely enough that most of them are of mixed ancestry, and that all, or nearly all of them, have received from time to time some infusion of Oriental or Arab blood. That would appear to apply in particular to the Welsh pony, who is on the whole the best-looking and most attractive of them all. On account of the inaccessibility of their islands of origin, it is equally safe to assume that the Shetland pony is probably the breed that has remained more true to its original type than any other.

Most of these pony breeds have their own breed societies and their own studbooks, and whilst it would lead us too far to try to describe each individual type in detail, it should yet be recognized that each one of these breeds has managed to capture the fancy of a certain number of enthusiasts, who have succeeded in carrying on the breed in which they are interested, keeping it very true to type, and in producing from almost all of them ponies of quite delightful quality.

Many of the smaller types, Shetland, New Forest, Exmoor and Dartmoor, have been bred as pit ponies, but some Forest and Moorland ponies are exceedingly nice riding animals for small children. The Shetland is a delightful toy, but is not on the whole really suitable as a riding pony. Welsh ponies produce beautiful riding animals for the bigger children and, besides, driving-ponies of quite exceptional quality and with much action. The cross of a Welsh pony mare with an Arab stallion will produce a riding pony of the very highest class.

The Highland pony has a number of enthusiastic admirers, and quite deservedly so. For, although the Highland is in truth neither a riding nor a driving pony, but more a utility pony suitable for land work on the hill farms of Scotland, or a shooting pony, there can be no doubt that it is very attractive in its own way, with a good bold head-carriage and an exceptionally plentiful adornment of flowing mane and tail of great beauty.

The Dale ponies are natives of Yorkshire and Northern England; they are miniature carthorses.

The Fell ponies are another heavy pony breed, found in the moorland districts of Northumberland and Cumberland.

Most, though not all, of the smaller and lighter types of horse belong to the warm-blooded variety and are predominantly of Oriental origin. By contrast the so-called heavy horses are predominantly of Western origin, descendants of the original breeds of North-Western Europe, and they form the cold-blooded variety.

The cold-blooded horses are of much heavier build than the warm-blooded breeds, and usually of greater size; their bones are bigger, though not so compact, and of lesser quality than those of their warm-blooded cousins; they are usually longer in the back; their quarters are frequently split, that is, with a clearly visible dividing depression; they are usually sloping, with a low tail-setting; their heads are big and coarse compared with those of the Oriental horse, with a comparatively long face and a small forehead, whereas the best type of Oriental horse has a small face and a big forehead. The cold-blooded horse is essentially slow and ponderous, but immensely powerful and pre-eminently suitable for heavy draught-work.

There is no doubt that our present breeds descend from the great war-horses of the Middle Ages, who needed to be of exceptional power in order to carry the enormous weight of their own and their rider's armour.

The main breeds of English cold-blooded horses are the Shire, the Clydesdale and the Suffolk, but the Percheron, which is a more recent introduction from France, deserves to be mentioned also.

The Shire is probably the best known and the most numerous of these heavy breeds. Undoubtedly, it is a very old breed, and it is generally assumed that their ancestry goes back at least to the twelfth century, when King John is known to have imported a hundred heavy stallions from the Low Countries which were famed for the quality of their great war-horses. From these early days the breed has gradually been developed into the great horse of today, which combines immense size, weight and power, with a very great deal of undeniable quality.

The modern Shire horse stands 17 hands high, or more, and

weighs about a ton; he is an imposing animal, on massive legs, with 11 to 11½ inches of bone in front, and up to 13 inches of bone underneath the hocks; the fetlocks are at a good angle, and his feet are wide and open; he has plenty of hair all over, long but not woolly, and the long feather to his lower legs, white more often than not, is of course characteristic. The Shire is a good, square and straight mover with good, strong and active hocks that do not turn in or out.

The modern Shire owes much of his quality to the activity of the Shire Horse Society, whose studbook goes back to 1720, to a horse called Blaze, bred in Leicestershire, a county which shares with Derbyshire the honour of having provided the bulk of the stallions entered in Volume I of the studbook.

The Clydesdale is the great horse of Scotland, quite as massive and as powerful as the Shire. There seems no doubt that both breeds descend from a common ancestry and that they are to a certain extent inter-related. But with the Clydesdale, the influence of Flemish stallions has been predominant in forming the breed. It is known that Flemish stallions were imported to Scotland during the Stuart reigns. The more immediate ancestry, however, is reckoned to date from the black stallions of Lochlyoch, imported by the Duke of Hamilton towards the middle of the seventeenth century, and standing at stud with one John Paterson. About a century later, the sixth Duke of Hamilton imported another Flemish stallion, this time a dark brown. It is known that for a long time afterwards the Lochlyoch mares were particularly famed and that most of them were blacks or browns.

The first stallion of the Clydesdale studbook is a horse called Glancer, also known as Thompson's black horse; he was foaled in 1836, and through his dam, a direct descendant of the Lochlyoch family, he gained a great reputation.

Though there is little to choose between the Clydesdale and the Shire in the matter of size and power, and though the Shire is himself a beautiful big horse, there can be no doubt but that the Clydesdale is the more magnificent of the two, and excels in quality and carriage.

The Clydesdale carries a very good head, with a broad

forehead and big eyes, and he carries it high and proudly on a beautifully-curved neck. Back and loins are short and broad, the quarters are broad and somewhat sloping, but with a good tail-setting. The hindlegs are powerful with faultless hocks. The breast is wide and supported on perfectly perpendicular forelegs, with big flat knees. Knees and hocks are close to the ground, fetlocks long and sloping, feet very wide and open, and of excellent quality. The Clydesdale has a good sloping shoulder with well-marked withers. It is important that his action be perfectly straight, and any tendency of the hocks to turn out is considered a great fault. The legs carry the same long feather as those of the Shire, but the Clydesdale usually has four white legs and more and bigger white markings than are as a rule met with in the Shire.

The breed is undoubtedly flourishing in Scotland, where many teams of these magnificent horses may still be seen in daily use.

The Suffolk Punch is the native breed of the Eastern Counties; it is remarkable in more ways than one. Though there is not a great deal of detail known about the earliest history of the breed, it seems incontestable that these horses, whose conformity to type is so striking, are descended from a native type of horse, a breed of long standing, with very firmly established characteristics. It is said that this breed owes little or nothing to the admixture of Flemish or other foreign blood. This theory is rather confirmed by the known fact that whenever stallions foreign to the native breed have been used, such as Blake's Farmer, a Lincolnshire-bred trotter, and Wright Farmer's Glory, also known as the Attleboro' Horse, their descendants failed to establish themselves in the long run as a lasting influence on the breed, but died out. It is undeniable that the original native blood has always prevailed, which is no doubt the explanation of the exceptional uniformity of type and characteristics in these horses.

The modern Suffolk is always a chestnut, with very little white; many shades of chestnut are admissible, but the reddish-chestnut is the most appraised. His temperament is lively, but very docile, and his courage and endurance are great; he will stand up to long days of hard work, is sober, and much less exacting in his nutritive requirements than the much bigger Shires and Clydesdales. The real Suffolk, though not a heavy horse, possesses plenty of strength, and is an excellent and very active farm horse. He stands somewhat over 16 hands, on very short, strong legs that are free from feather, he is deep in body, with a wide breast and strong quarters. The head is good, with a straight profile, a good forehead and large eyes, set on a rather short and heavy neck; the withers are not well marked and the shoulder is rather straight and fleshy.

The Percheron is one of the most beautiful heavy horses, and it is not therefore surprising that this breed, introduced to this country from France, has found so many staunch supporters in England.

The prototype of the breed, which originates from La Perche, is the Petit Percheron or Percheron Postier, an active, very powerful horse with much quality, standing about 15½ to 16 hands. He was a good-looking horse, with a small head, wide nostrils and a big, clear eye; neck somewhat short, but beautifully curved, with a long silky mane; a good but not a prominent wither, the shoulder somewhat round, but very muscular; legs excellent and very muscular, with little hair; a strong-made breast and exceptional quarters, that were described as having *une expression harmonieuse de la force, d'une certaine vitesse et de la résistance.* Clearly, some of these characteristics must be — as they were — due to the influence of Arab blood. As the name indicates, the Percheron Postier was a coach horse, and he was appreciated also as a weight-carrying riding horse.

His type was immortalized in the painting by Rosa Bonheur that attracted the admiration of the Empress Eugénie and became the success of the Paris Salon of 1853.

Gradually, under the influence of the ever-growing demand for heavy horses, especially from America, the *Petit* Percheron gave way to *Le gros* Percheron, the modern, big

66

Percheron horse, who stands from 16 to 17 hands high. Though the big breed as we know it today is still an exceedingly fine type of heavy horse, it is yet undeniable that much of the original quality was lost in the process of getting the additional size and weight. The head, originally fine and almost typically Arab, became heavy, big, and somewhat coarse. The neck is now very massive, but of adequate length, withers are low, shoulders sloping but fleshy. The quarters are immensely strong, long and broad, somewhat sloping, but with a high tail-setting. Usually the quarters show the division or split that is so typical of the cold-blooded breeds. Legs are strong and muscular, though the second thigh above the hocks is not always perfect. Feet are good, small rather than big.

These great horses are capable of the heaviest work, their temperament is somewhat lively, they are high-couraged and active, and fast in their work.

The true Percheron is born black and gradually turns a beautiful dapple-grey, the colour being at its best at about 7 to 8 years old; in later life they turn lighter, and finally may become almost white. Undoubtedly this colour is still an inheritance of the original admixture of Arab blood.

CHAPTER FIVE

Reflections on Breeding

*Heredity—Linebreeding—Breed   Improvement—Affinity—Mating
Selection—Judgment—Leading   Characteristics—Temperament—Action
—Conformation—Constitution—Soundness—Performance—The   Old
Favourite.*

ALL BREEDING begins with the mating of a female to a male.
Which is very simple, and at the same time very mysterious.
There is a good deal about it that we do know, probably
because it is so simple, and there is a good deal more that we
do not know, probably because it is so mysterious.

There is no such thing as a law, or a set of rules, or even a
concrete science that we can apply to breeding with any
certainty.

But, provided we bear this limitation in mind, there is a
great deal that we can learn from the experience of generations of able breeders, the living proof of whose success is
with us in the very excellent horses that grace the best of our
modern studs.

Experience and observation have brought us much
valuable knowledge regarding the results that we may
expect with some confidence, though never with certainty,
from certain principles of breeding; they have even brought
us careful theories worked out into much detail, particularly
in connection with the breeding of thoroughbred horses, as
for instance the theories of Bruce Lowe and of Colonel
Vuillier.

It is left to every breeder's common sense to borrow from
the existing reservoir of experience, observation and knowledge, the precise facts that appear to apply most nearly to his
particular case or problem.

To do so really well is an art. For to do so, implies the

ability to see in one's mind's eye the produce that is likely to result from the mating of one individual with another, and, moreover, to see the qualities and the defects that the produce is likely to have.

To do that, implies more than the mere ability to recognize both the qualities and the defects in each individual parent; for though that in itself is difficult enough, the real breeder must visualise how the interplay of these qualities and defects is likely to affect the offspring.

Though parents will, as a rule, transmit their main characteristics to the foal, that rule is subject to several limitations. For the parents themselves are products of inheritance in whom their parents' or even their ancestors' characteristics are prevalent, in some cases markedly so, and in some other cases lying dormant. The transmission of certain characteristics may jump one or more, or even a number, of generations, and in that way it happens that characteristics may come out in the foal that had been quite invisible, though certainly present, in one of the parents. Such an occurrence is called a throw-back, and happens, for instance, when a foal of two nice and apparently well-bred parents suddenly appears with hairy heels or with a common head, when it has obviously thrown back to some common ancestry of one of the parents, usually, of course, of the mare.

Obviously, if both parents are well bred, and not only look it, such detrimental throw-backs will not occur, although it still remains true that both parents will be transmitting the characteristics they have themselves inherited from their forbears.

That is where the value of Thoroughbred stock comes in, because we know that in such stock leading characteristic qualities have become more and more certainly fixed by generations of carefully selected breeding.

That is, too, why breeders who wish to develop one particular characteristic more than any other, as for instance speed or stamina in the case of racehorses, or milk or high weight gains in the case of pedigree cattle, will go further still than the selection of Thoroughbred blood, in that they will endeavour to breed only from a few selected lines of blood wherein these particular qualities have been especially marked. This is then called line-breeding.

The breeder of racehorses should undoubtedly study the theories of line-breeding that have been advanced by Bruce Lowe and the others, in the special literature that exists on the subject. Since this book is not intended for the breeders of racehorses in particular or, for that matter, of any other particular breed or kind of horse, these theories are beyond its scope, apart from being too vast a subject in themselves, to be susceptible of being dealt with in an abbreviated form.

Here the subject has been mentioned only to accentuate two things. Firstly, that in breeding from a certain sire and from a certain dam, we are in reality breeding from that dam's and from that sire's ancestry. And secondly that, when considering the points of any horse from the breeding point of view, we shall do well to remember that the better such a horse is bred, the more surely are his characteristics likely to be fixed in him, and the more likely he will be to transmit them to his descendants.

The stallion and the mare are both equally potent in stamping their type and make-up upon the foal that is to be the product of their mating, and the foal will therefore be a blend of its parents' qualities and defects. But the equal potency of both parents applies only in so far as both are equally well bred, for if one or other of the parents is better bred than the other, the better-bred one will be markedly pre-potent. Which is due, of course, to the accumulated influence of a flawless ancestry.

That is the leading principle whereon all breed improvement is based and without which no breed improvement would be possible. The influence of any high-class stallion upon the improvement of an inferior breed may be quite enormous, and in that respect the English Thoroughbred is very potent. He has been used for that very purpose with immense success the world over. In France that very good breed *le cheval Anglo-Normand,* has been created by the free admixture of Thoroughbred blood with the original native stock of Normandy horses. At the Haras du Pin, and at other great State-owned studs of France, Thoroughbred stallions of high class are kept for the very purpose of improving the native horses. Other famous studs where Thoroughbred stallions have been used for the same reason are Trakehnen, Babolna, Kisber, Mezöhegyes, and there are many others.

The potency of the Arab in this respect is greater still; his blood, as we have seen, permeates that of the Thoroughbred himself, and, besides, has been the fountain-head wherefrom the best characteristics of all breeds of warm-blooded horses derive. The Arab carries, compressed as it were in his small body, the greatest possible array of equine qualities, and, moreover, these qualities, by centuries of pure breeding, have become so firmly fixed that his potency to infuse them into other breeds is almost beyond belief.

As an illustration we may refer to the Alcock Arabian who is, as mentioned by Lady Wentworth, the ancestor of all grey thoroughbreds now living, the colour having persisted in unbroken descent for 240 years.

My own stallion, Basa, who is of the Bablona Stud's Shagya breed, is a grey, with a pedigree extending back to 1798, covering thirteen generations; his ancestors were predominantly white or grey horses. This horse, who is now 16 years old, has had a very successful stud career and has served a large number of mares, as many as forty in a season. Though the mares have been of all colours, and all types of breeding, I have yet to hear of one single offspring that did not turn out a grey. Some of these mares, sent to me every season by their owners, have been of very poor quality and, truthfully, unpromising as breeders; there have been, besides Arabs, Thoroughbreds, big hunter mares, small pony mares, and even cart mares of the smaller type. And yet, though the produce has, of course, varied in conformation, every single foal has had the horse's mark stamped upon him so unmistakably that one could never fail to recognize them anywhere as the produce of the same sire.

It is well known that the Arab is much less liable to any of the many hereditary unsoundnesses that are so troublesome in other breeds, including diseases of the bone and, of course, wind. The Arab's bone is especially hard, compact and dense, and any bony defects are almost unknown; the Arab's wind is always perfect, and whistlers, roarers or broken-winded Arabs are unheard of. I have it on the authority of the great State studs in Hungary, where Arab sires are used constantly for the improvement of native breeds, that one single infusion of Arab blood has often been found sufficient to outbreed prevalent types of native unsoundness of the kind that I have just mentioned.

But, however beneficial the use of Thoroughbred and Arab blood may be for the improvement of native breeds, and equally for the purpose of breeding individual produce from mares of less refined breeding, there is one limitation that should not be lost sight of. There must be a reasonable affinity between the breeds or types to be crossed.

Do, by all means, put a hunter mare or any type of half-bred mare, a good pony mare, a hackney mare, a trotting mare or even a Cleveland Bay to a Thoroughbred or an Arab stallion and be confident of good results. But don't expect either of these highly-bred horses to produce reliable or durable results from a cold-blooded cart mare, with whom they have no affinity.

I have seen this crossing of Thoroughbred or Arab stallions with cart mares recommended lightheartedly as a means of breeding heavyweight hunters; I know that wonders do occur, and it may be that an occasional crossing of that type will yield a good result, though I have never seen it. But for one success, there are bound to be a hundred failures. These failures will be neither hunters nor even improved cart-horses, but on the contrary, plain mongrels. The same applies of course to the mating of Thoroughbred or Arab mares to carthorse stallions.

The ancestry of the horses from which we intend to breed is of great importance therefore, because good ancestry enhances the likelihood of such horses passing on their good points. In well-bred horses good points are no flash-in-the-pan, as they are sometimes in less well-bred animals. We may even go farther than that, since good breeding will often outweigh minor defects in conformation; the chances are that the animal's ancestry may enable him not to pass on such defects; no such likelihood exists in the case of animals of more common breeding.

However, unless we know what our animal's ancestors have been like, and what they have been able to perform, as we shall know in the case of high-class racing stock, the knowledge that our animal is a well-bred one is largely a theoretical consideration, though an important one, with which to back up our judgment.

But that judgment itself, in practical form, can be directly concerned only with the actual animals wherefrom we propose to breed. Now, points that are similar in both parents are likely to be confirmed in the foal or even accentuated; so two animals with good heads are almost certain to produce a foal with a good head, and probably one with an exceedingly good one; but, on the other hand, two parents with bad heads can be relied upon to produce a bad head, and more than likely a very bad head, in the foal. If one parent is poor in one particular point, there is a very reasonable chance of improving upon it in the foal, provided that the same points are particularly good in the other parent.

Now it is nicest, of course, to be able to match two parents with nothing but good points matching each other, in which case we may be pretty certain of producing a first-class foal. Unfortunately, such a combination of perfections is so rare as to be almost impossible.

The next best thing, and in practice about the only thing that we may hope to achieve, is to match our horses in such a way that the stallion will be particularly good in points wherein the mare is deficient, and vice versa. The worst thing to do is to match two horses that fail to any marked degree in identical respects.

It follows that in order to be a successful breeder one must also be a judge of a horse, not necessarily in the sense of being able to judge horses in the show-ring, but, most definitely, in the sense of being able to appreciate the relative importance

73

of certain shortcomings in our stock, having regard to the purpose whereto their produce will be put. A good many people are quite capable, and often quick enough, to spot shortcomings in other people's horses, whilst they appear genuinely unable to spot the weaknesses in their own animals. And that is about the most serious defect in any breeder's armour.

We must remember that no such thing as a perfect horse exists; every animal, however great his class and quality, is bound to fail to come up to the ideal in some respect or other, however slightly that may be. In fact, the failing may be so slight that it might escape many an experienced eye during a cursory examination; but it should not escape the eye of his owner, who sees his animals every day, under all sorts of conditions, in every kind of attitude, alert and sleepy, fat and thin, polished and rough, at their best and at their worst. It is up to him to study his animals constantly, carefully, and above all with an unprejudiced mind that is open to recognize and to admit shortcomings as well as qualities. It is in fact more important to recognize the former than the latter, for it is only so that we shall be able to make a wise and reasonable breeding selection.

From a breeding point of view all the leading characteristics of the sire and of the dam are hereditary: that is temperament, action, conformation, constitution and soundness, and — as a result of them — performance.

I place temperament first, because for most purposes even the most perfect horse will be greatly marred if his temperament be unsuitable, whereas actual vice may render him almost or even entirely useless. This applies to all horses, no matter for what purpose they are intended, but it applies particularly to horses intended for some form of personal use. It does not, perhaps, matter quite so much in the case of racehorses, when speed is the all-predominant requirement, but even racehorses that are too temperamental are seldom successful.

Naturally, the exact demands that we are entitled to make on temperament vary within very wide limits from one breed or type of horse to another. Whereas we want to see courage and grit in all types of horses, we should no doubt condemn the highly-strung temperament that is normal in a horse with much blood, such as a Thoroughbred or an Arab, if it were met with in a carthorse; we should do so because a hot carthorse would not be very suitable for his job, which requires a placid animal. Likewise, although we cannot have true blood and quality without high, and sometimes even exuberant courage, we do not want too hot, and certainly not too gassy a temperament, in a hunter, since horses so endowed are too difficult to manage, and may even be impossible in company. Manners are very important.

But, no matter what type of horse we are dealing with, we are entitled to look for kindness and confidence, since these two traits will, in suitable surroundings, go a very long way towards making up for excessive courage.

It is well known that mares are frequently very highly strung, but, provided they are also kind and gentle, as is so often the case, their nervousness need not deter us from breeding from them. But it would not be wise to put them to a stallion who is himself of a nervous disposition; on the contrary, we must, for such mares, look for a stallion with a particularly kind, quiet and reliable temperament. In that way we may influence the temperament of the offspring a great deal.

But not in that way alone. For it must here be recognized that temperament, though so largely influenced by heredity, is most definitely influenced also by environment and example. By taking more than normal interest in a nervous foal, handling him especially gently, and placing him in the company of some other foal that is itself quiet and confident, much can be done to allay his inborn fear and nervousness.

Vicious horses are fortunately rare, and horses whose vice is inborn, and therefore hereditary, are rarer still. The bulk of vicious horses met with have become so as the result of injudicious, rough or cruel treatment. None the less, real viciousness is a frightful thing, and in my opinion vicious horses should not be bred from.

75

I place action second, because, however good he may appear in other respects when standing, a horse that cannot move well and truly is but a poor specimen. Moving well implies that liberty and scope of action that have been referred to already in the previous chapter. If a mare is deficient in this respect she should be sent to a stallion who is himself a brilliant mover. True action is straight action, without dishing; and level, without lameness. Slight defects in straightness of action may be overlooked, provided again that the stallion be irreproachable in this respect. Lameness is of no importance when due to an accident, but is a grave defect if due to constitutional causes such as debility or rickets, or other causes predisposing to unsoundness. As to that, more on a later page.

Conformation is very important, because whatever we breed we all want good looks for a number of reasons. Good looks, in the first place, are a pleasure in themselves, and enhance the enjoyment to be derived from our animals, and the justifiable pride to be taken in them. Good looks have a very great influence on the value of a horse. Finally, good looks are more often than not the visible expression of excellence.

In judging conformation we shall never go far, in my opinion, if we divide the horse into a catalogue of points and try to assess all those individually; neither does a method appeal to me which I have seen advocated with proverbial German thoroughness, whereby the horse is reduced as it were to a geometrical problem, measured up in all his various component dimensions with a yardstick, and weighed. Conformation, again in my opinion, can only be judged effectively by looking at the picture the animal presents, both standing and in movement, and if that picture is a real eyeful of quality and balance, there is probably not a great deal wrong with him or her. I look for quality first, see the animal as he approaches me or as I approach him, head, front and shoulders, expression and bearing. These characteristics are unmistakable and can be seen at a glance, and if they don't please me I have no interest in considering the animal further. But if the first impression is pleasing, I will

76

look him over in more detail and will try to take in all the good points first, breast, depth of body, loins and quarters, tail-setting and carriage, limbs and feet. No horse is perfect, and it is therefore useless to expect one without short-comings or faults. If we look for faults first, we are making a wrong mental approach, and are liable to condemn a good animal merely because we are fault-conscious for some shortcoming that may be relatively unimportant. For the importance of most shortcomings is only relative, inasmuch as it depends primarily on the precise purpose for which we are breeding. As an instance, the best quarters should be long, level and broad; sloping quarters, in a riding horse, are certainly against him as a point of conformation, and would weight pretty heavily in the show-ring; but provided these sloping quarters are of appropriate length and broad and muscular, they will not affect the horse's performance either as a hunter or as a racehorse; in fact, it is noted that such horses are frequently outstanding jumpers. But sloping quarters that are at the same time short, render a horse unsuitable as a riding horse, although they are no detriment in a carthorse. And if, in addition to being short, they are also narrow and therefore weak, they become indeed a serious fault in any type of horse.

Similarly, we all like a straight hindleg, and sickle-hocks are certainly no show point; but they are no detriment to speed or performance, and many first-class thoroughbreds have them; they need be regarded with circumspection only in cases where the hocks are also narrow and weak, as in that case they are liable to lead to trouble.

The same thing applies again to horses standing over at the knees. Though not exactly pretty, it is, if the animal is born that way and the disfigurement, if it may be so called, is not due to the effects of work, a sign of strength rather than of weakness. Such horses will very seldom break down. Conversely, standing back at the knee, which is less easy to notice and far more prevalent than standing over at the knee, is a much more serious fault since it is a sign of weakness in the limb.

A horse with pintoes, that is toes turned inwards, is never

liable to strike himself, and therefore pintoes are much less serious as a defect than toes turned outwards, which are liable to make him do so, and even to bring him down.

These and similar shortcomings are relative in the sense explained above, but they are relative also in the sense that most or all of them can be bred out in the foal, or at any rate vastly improved, by the judicious choice of a stallion who shows excellence in the very points where the mare shows deficiency.

Now a good brood mare, though bound to have some weak points, since no living creature is without any, will not have very many defects to guard against, and it will be comparatively easy to find a stallion that will compensate for her shortcomings. But, however excellent stallions may be, they too will be found to have their shortcomings. And it follows therefrom that it will become utterly impossible to find a stallion perfect enough to engender a first-class foal from any mare that is, as some are, a walking exhibition of defects in conformation. In other words, whilst we need not be unduly perturbed though our mare can be crabbed on one or two points, if she is otherwise a good one, we had better not attempt breeding from any mare that is full of faulty points.

Flat-ribbed, narrow-breasted, shallow or herring-gutted animals have not the proper room for heart and lungs, and those are very serious faults since they affect constitution as well as conformation. In a mare these faults are quite unpardonable, since they deprive her of the necessary room wherein to carry and to develop her foal. Brood mares must be wide and roomy, which means well ribbed up and broad and muscular over loins and quarters.

Colour, though not really a point of conformation, is yet one that affects a horse's looks, and is, therefore, not without importance. People are sometimes inclined to very strong preferences for certain colours, mostly on the ground that they are either flashy, such as chestnuts, or distinguishing, such as greys. In a sense that is a little childish, since colour has undoubtedly very little, if anything, to do with a horse's quality. On the whole, we may assume that any colour is good, provided it is clear and definite, with some depth to it, and not washy or mealy. Pure black horses are rare; most black-looking horses have a brown nose when, since a horse's true colour marks from his nose, they are called brown. A good deep brown, almost black, is much favoured and a hardy colour. A real rich bay, usually with black mane and tail and black legs, and sometimes black points, takes a lot of beating, and I am not sure that it is not the best colour of all. Pure white, from birth, is exceedingly rare, though greys will often go pure white, or nearly so, as they become older; greys are usually born black, or nearly so, though they may be born almost any colour, and the change occurs very gradually; greys are at their best at about 6 to 7 years old, when the colouring with its rich dappling and many other variations may be very lovely; the colour, to be good, should not look faded or with a dusty shade over it. Chestnuts exist in a great variety of shades, from golden and red to liver-chestnut; these colours are popular and attractive, as long as they are not washy, which is ugly and often denotes lack of quality. Cream is a fairly rare colour in English breeds, and is usually combined with a lack of pigment in the skin and eyes; it is apparently a mild form of albinism. Duns have a blue or yellow shine over their coats, with a black skin. Roans have a mixture of white hairs distributed all over the true colour of their body. Piebalds are marked with large and clearly defined black and white patches, and skewbalds with similar white patches combined with those of any colour other than black. Odd-coloured horses are those with patches of more than two colours that tend to merge into each other.

Professor Anderson, of the Kentucky Experiment Station, carried out considerable investigations on the subject of colour in horses. He found that the chestnut colour is recessive to all other colours, and that a chestnut horse can produce only one kind of reproductive cell as to colour. Consequently, when a chestnut mare is bred to a chestnut stallion the foal will always be a chestnut. That is the reason why it was such a comparatively simple matter to fix the chestnut colour in the Suffolks.

But with horses of all other colours it is impossible to

predict, with certainty, what the colour of the foal will be.

Constitution is the next important factor to take into consideration, since we want strong, healthy and thriving stock. As I have already pointed out, certain defects in conformation, such as narrow chests and flat-sided ribs, imply that the animal's constitution cannot be first class, since there is insufficient heart and lung room. But apart from such defective conformation, as is pretty obvious, there are always individuals who, though well enough made, are not strong in health and lack in stamina. The case may be one of general debility, possibly inbred or else the result of lack of care in early life, or perhaps a consequence of some serious illness or infection. It is met with fairly frequently in thoroughbred stock. Such horses will not thrive properly, they will look poor when others look quite well, they will recover very slowly from any setback, and will stand up badly to inclement weather or to any other form of hardship. Unless very well fed, they may be dull and lifeless. They may be dull in the coat, and in many respects look very similar to cases of serious redworm infestation.

But, if redworm is not present, which can be checked by a veterinary examination of the droppings, we may be sure that the trouble is constitutional. Horses so affected will almost always breed weakly foals that are really not worth rearing. Therefore they should on no account be bred from.

Though I have left soundness to the last, it is by no means the least important consideration, in fact it may well be the most important one, because an unsound horse is no good to anyone, and because so many forms of unsoundness, or rather the tendency to contract them, are very hereditary. That does not apply, of course, to unsoundness as the result of an accident.

There are various forms of unsoundness which the prospective brood mare may have contracted after birth, such as roaring, navicular disease, sidebones, curbs, ringbone and spavin. Such forms of unsoundness cannot and will not be transmitted by her to the foal, who will not be born with any of these defects. But what the mare quite possibly may transmit is a tendency to contract such infirmities, owing to weakness or malformation of certain organs or joints.

Since the question as to what unsoundness is, or is not, hereditary, is subject to some controversy, I will not attempt to give a comprehensive list. I will only say, but that with considerable emphasis, that any kind of unsoundness in a studhorse should always be viewed with considerable suspicion, and that applies to the following in particular: any kind of defective wind or eyesight, spavin, curb, any diseases of the foot, and any bony diseases.

In many cases the seriousness of the complaint, from a breeding point of view, depends upon its exact nature and whether it appears to be due to any malformation of the affected part, and also as to whether or not the complaint, if passed on to the offspring, is likely to affect that offspring's usefulness.

The only way to obtain sound advice in such a case is to call in a veterinary surgeon who has made a speciality of stud-work, and preferably one who is, or has been, a horseman himself. Such a one will have, besides the advantage of his training, a far wider, deeper and more reasoned experience than any private breeder or stud-groom can ever acquire.

One should be very reluctant to breed from any unsound horse.

Performance is, as I have said, the resultant of temperament, action, conformation, constitution and soundness.

Performance is the manner wherein our horses acquit themselves of the task for which we require them, and its importance is equally great no matter what may be the exact nature of the task, whether on the racecourse, in the show-ring, in the hunting field, at the plough, or even in teaching our small children to ride. A horse that is sound, of strong constitution, with good conformation, good action and a nice temperament, is bound to acquit himself well, and probably supremely well. And, since all these characteristics are hereditary, it follows that performance itself, which is the culmination of them all, is also hereditary.

That is why performance is such a very important

consideration in the breeding of racehorses. And that is why
it should be, whenever possible, an important consideration
also in the breeding of all other types of horses. A mare that
has proved herself a good and reliable farmworker in all
gears, is likely to get one that will be the same. The same
applies to the hunter, or the child's pony, or every other kind
of horse. If we know that the dam was a good one in her
work, and that the stallion was so too, we may feel pretty
safe that the offspring will not disappoint.

The above are the main reflections on breeding that come
to my mind as essential points to be considered by any
breeder who is intent upon making a success of his industry.
Breeding is a long-time job with a great deal of work
attached to it, and with very considerable expense involved.
It costs no more to keep a good mare than it does to keep a
bad one; and the cost of rearing a poor or moderate foal to
maturity is no less than that of rearing a good one. And in that
way, to be penny wise in breeding, is most certainly pound
foolish.

Though the above is very true, I yet want to say a few
words about the more or less casual breeder, who decides to
breed a foal or two from an old favourite. I have seen this
practice, which is much more widespread than many people
imagine, deprecated on the grounds that the mare lacks
sufficient quality or else that she is too old. Though it is true
that a mare without much quality is not likely to breed a top-
class animal, there is yet a lot in the fact that she has been able
to capture her owner's affection to such an extent. Quite
certainly, she must be a kind and likeable sort, with the right
sort of temperament, and more likely than not she has
performed to her owner's entire satisfaction. These are,
surely, two very important recommendations that cannot be
brushed aside lightly. If she has answered her affectionate
owner's requirements so satisfactorily, then I think it is more
than likely that she may produce a foal that will indeed
perpetuate both that satisfaction and the mutual ties.
Particularly so, if she can be sent to the right sort of stallion.

And as regards being too old, well it is certainly more difficult to get a mare of 12 or more years old, that has not bred before, in foal than is the case with younger mares; there is definitely an increased risk of barrenness and of a wasted stud fee. But that risk is by no means as great as it is sometimes made out to be, and I know of cases of mares of 20 years old and more, that have bred most creditable foals at the first time of trying.

If I had such a favourite mare, I would most definitely try to breed from her, even though I felt quite certain that the foal would not be out of the top drawer.

CHAPTER SIX

Sending a Mare to Stud

Seeing the Stallion and his Stock—Limited Number of Approved Mares—
Maiden Mares—Difficult Breeders—Stud Fees, Groom's Fees, Keep of
Mare—Hunter Breeding—Premium Stallions—Fertility—Barrenness—
Condition—Travelling—Time Required at Stud—Best Time for Breeding
—Railway Tariffs—Stud Accounts.

IT IS HOPED that the considerations set out in the preceding
chapter will be found helpful in selecting a suitable stallion to
which to nominate our mare. But it will be borne in mind, of
course, that these considerations, which are largely
theoretical, cannot be put into practice very well unless we
take the trouble to go and see the horse.

The stallion owner, or his stud-groom, will be pleased to
get his horse out for you, so that you may see him to the best
advantage; but do not go without making an appointment —
stud farms are busy places, particularly during the breeding
season, and certain times of the day that are set aside for
trying or covering mares are quite unsuitable for dealing
with visitors.

When you have seen the horse, and you think you like him,
ask to see some of his stock; almost all studs breed from their
own stallions, and they are almost certain to have youngsters
about of varying ages. To see the stock is most important, in
fact more important in a sense than seeing the horse himself,
since it will show you what he is capable of, in the way of
leaving. Some very outstanding horses leave but moderate
stock, whereas some more moderate horses are at times
capable of leaving stock much superior to themselves. That is
but one more example of the fact that there is no such thing as
certainty in breeding. However, if the horse is of a type and
character that pleases you, and that seems to suit your mare,

84

and if, moreover, his stock are nice, level and full of quality, you are not likely to go wrong in fixing upon your nomination. In viewing the young stock one should look out particularly for the type that most high-class horses manage to impress upon their offspring. If that type be a good one, the horse must be a good sire; if, on the other hand the young stock are markedly defective in some special point or other, as for instance coarse heads, pig-eyes, poor fronts, bad hocks, or some similar failing, you may be pretty certain that the horse has a propensity to pass that particular weakness, although he may show no sign of it himself.

Whilst looking over the horse and his stock, it is as well to have a good look around generally; firstly, one may usually pick up some useful knowledge by looking around a well-run stud, and secondly one would like to know that the stud where one is going to send one's mare is indeed a well-run one. Sending a mare to stud is always a matter of some expense, and sometimes of considerable expense, and there is no doubt that efficient management and the general way wherein a place is run have a great deal to do with your mare's chances of being got in foal.

Mares are apt to be temperamental, and may vary a great deal in their breeding propensities, and to do them justice requires a great deal of careful and even minute supervision, much patience, much hard work, and a great deal of knowledge, experience and conscientious effort. You should feel satisfied that these requirements can and will be met at the stud of your selection.

When you have finished your inspection, you will not forget of course that you have taken quite a bit of the stud-groom's time, which he will have to make up later in the day; and stud-grooms are very human, so that a tangible expression of your appreciation is not likely to be taken amiss.

If it so happens that you are a breeder of racing stock, you will of course be guided also by the stallion's own performances, by those of his ancestry, and particularly by those of his stock if the horse of your choice is old enough to have produced stock to race. In that case, your choice will be limited to a comparatively small number of horses. It will be further limited by the fact that the best stallions have no difficulty at all in filling their lists, and that they are limited to a stated number of approved mares. The maximum number of mares that a stallion in full vigour can cover in a season, and do justice to his mares, should not exceed forty. If it is considered that the breeding season is compressed into a few short months, and that some mares may have to be covered two or three or even more times, and that occasionally more than one mare must be covered on the same day, it will be understood that forty is already a very large number. Since the reputation of a stallion, and his future success and earning power, are based to a considerable extent upon the horse's record of fertility, it is understandable that wise stallion owners are not prepared to overwork their horses. It is incidentally very much in your interest as the mare owner that this should be so, as otherwise the chance of getting your mare in foal will most certainly be affected.

Apart from fertility, it is the quality of the stock produced by a stallion that will affect his reputation more than anything. And, as the horse cannot do it all himself, he will only be able to show what he is capable of by being mated to good mares. Which is one reason why the owners of a good horse will only accept approved mares, that is mares that are likely to do their horse justice. Another reason is that some mares are more difficult to breed from than others, or more uncertain. That is why for popular stallions, for which it is easy to get a full list, owners will not as a rule accept maiden mares, and a good few will also refuse to take barren mares.

The proportion of mares whose fertility is sub-normal is quite considerable, and the most successful studs are not anxious to accept mares that might depress their horse's fertility record. Naturally, maiden mares are an unknown quantity in this respect, and moreover they are often much more troublesome to cover for the first time, so are seldom accepted to first season stallions or old horses. When a mare is a breeder and fully and properly in season, the moment will always come when she will take the horse willingly, and without attempting to kick. None the less, maiden mares are sometimes very nervous and fidgety, and may require a good deal of steadying and holding in place. That is why unbroken maiden mares are an abomination, since not under proper control, and no mare should ever be sent to the horse unbroken. But apart from steadying a young mare by legitimate means, constraint, in the form of hobbles or similar devices should never be used. As Sir M. E. Burrell points out very rightly in his excellent little book, *Light Horses, their Breeding and Management*, it is an indignity upon a mare to force her to submit to the horse, that she will never forget or forgive, and that may do untold harm. And it will certainly not do her any good, and will not get her in foal, since conception can only occur at the precise time when the mare is fully in use.

If a mare cannot be induced to take the horse willingly, her owner is entitled to be advised of the fact. It is usually worth while to have such mares examined by an experienced veterinary surgeon. At other times, mares whose sexual instincts are dormant may be brought into use by an injection, or by a series of injections, of hormone. There are also cases of maiden mares who, although well and properly in season, are yet so nervous and scared of the horse that it is a physical impossibility to cover them. In such cases, insemination may have to be resorted to, providing the breed society will recognise the resulting foal.

But all such unusual or suspicious cases ought to be dealt with under veterinary assistance, as one can never be quite certain without an examination as to what is wrong with such mares. Some time ago I was sent a favourite mare whose owner was most anxious to have her covered because, as he told me, she was almost perpetually in season. This mare was a maiden, and we took the usual almost endless trouble with her to get her to take an interest in the horse. Her behaviour was most peculiar; whilst she took no interest in the horse and showed no signs of coming into season, she did not rebut him either, as normal mares do when they are not in use, in fact she appeared completely unaffected by the horse's attentions; she seemed, in fact, completely lifeless. Upon my suggestion, the owner had this mare examined. It was found that she had no womb or breeding organs at all, and consequently no room to receive the horse. If that mare had been served by force the horse would have caused a fatal internal injury, haemorrhage, and certain death. An extreme and very rare case, no doubt, but still one that goes to show the sort of risk that may be incurred by forcible covering. Naturally, this mare could not come into season, and never had been, so what caused her owner's misapprehension it is difficult to say, excepting that from my experience many owners, and grooms also, are quite incapable of assessing just when a mare is in season.

On another occasion I was sent a nice blood-mare that had been involved in a traffic accident; she had been run into from behind and carried a large scar over her quarters, that had been sewn up and had healed well. Only, some contraction and deformation had taken place, and as a result the mare's vagina was no longer straight, but crooked. When I saw the mare I did not think she was fit to breed, and wrote and told the owner so. In reply, the owner informed me that she had taken veterinary advice on the matter, and that in her vet's opinion the mare could breed. So, on the strength of that assurance, we proceeded to take this mare to the horse and attempted to cover her when she came fully into use, as she undoubtedly did. Now this was a gentle, mannered and obedient little mare, but when the horse approached her she became quite unmanageable and kicked and plunged like fury. In my opinion, that mare knew quite well that she must not breed, since she was all contracted behind and would no doubt have had a frightful time foaling, even if she had been able to take the horse, which seemed doubtful.

These examples may illustrate that you are better off when a genuine stud returns your mare unserved, with no fee to pay, than when a not-so-genuine establishment uses force merely for the purpose of making you pay that fee.

However, to retrace our steps, upon your being satisfied with the stallion and his stock, and subject to the stallion owner's approval of the mare that you propose to send, you can now make your nomination. You will be told, if you do not already know, the amount of the horse's fee, that of the groom's fee, and you should be told, or enquire, the charge made for keep of mares, also the swabs your mare will need before she can be accepted onto the stud or covered.

The stallion's fee depends entirely upon the value placed by breeders upon his services. In the case of leading racing sires, the fee of the best and of the most fashionable stallions, which is usually though not necessarily the same, is very high, and may be anything up to £100,000 or perhaps even more. Fees of £20,000, 10,000 and 5,000 and around 2,000 guineas are common enough, and stallions standing at below the latter figure are cheap and can hardly claim consideration as fashionable racing sires. Since it is in reality the law of supply and demand that settles the amount of the fee obtainable for a horse's service, it follows that the most expensive horses are usually those that are most in demand, and to whom it is the most difficult to secure a nomination.

Apart from these expensive stallions, whose services are sought after mainly for the breeding of prospective race-horses on the flat, there are usually a number of much cheaper horses who, without having shown brilliant speed, may have been good stayers or good jumpers, and are in demand for the breeding of prospective jumping stock. Some of those make excellent hunter sires. Their fees would range from about £150 to £500, although some horses will serve mares for as little as £80.

The groom's fee varies in sympathy with the horse's fee, and is usually from £5 to £20. When an inclusive fee is quoted for the horse, that means that the horse's owner will discharge the groom's fee himself.

Terms for keep of mares do not vary a great deal at the well-established studs, and may be from £35 to £77 per week for barren mares kept at grass, and from £49 to £91 for foaling mares that are brought in at night and are corn- and hay-fed.

Novice breeders sometimes appear to think that these terms of keep are rather excessive. Actually they are nothing of the sort and, if anything, they are too cheap to be a paying proposition. A well-organized stud, well-staffed with experienced men, with good paddocks, well fenced, and with plenty of grazing and other good keep, is a most expensive place to run. There is a lot of work attached to supervising all these visiting mares, and anything in the region of forty mares is a full-time job for at least two good men. Incidentally, forty or fifty mares, staying about ten weeks each, get through an enormous amount of grazing and usually leave the paddocks so bare behind them that recovery will not set in until the autumn. Those paddocks might have produced 40 or 50 tons of valuable hay, had it not been for these visiting mares.

With regard to the horse's fee, one must decide in accordance with the value of one's mare, with the type and the reputation of the sire, and with the likely value of the produce. Whereas it would naturally be ridiculous to send a hunter mare to a £10,000 stallion, even if that could be done, it would be equally foolish to send her to an indifferent £50 horse, if a much better sire could be found at £80, £100 or even £150. The £25 or £50 saved in this way may well make a difference of two or three hundred pounds, or more, in the ultimate value of the foal. For the best type of hunter, too, will always command a big price, and the amount of the stud fee will be no more than a fraction of the cost and trouble of rearing the foal to maturity.

Those who breed thoroughbred or pure-bred stock will naturally send their mares to a stallion of the same breed, and in a sense that will help to narrow down the field of their choice.

This applies in particular also to Arab breeders, although the number of Arab studs is very much on the increase. The best of these stallions are in a comparatively few hands, and although less expensive than the best Thoroughbred racing stallions, a good Arab may command quite a substantial fee. From £100 to £500 is a reasonable fee for a good horse of the best blood, but some stallions stand at up to £1,000. Still, there are some quite good Arabs about, suitable as potential pony, hack, or even hunter sires, that stand at a fee of £100.

Though it may be difficult enough to breed the best kind of pure-bred horses of any breed, I yet believe that the man who sets out to breed a cross-bred type of horse, as for instance a hunter, has a more difficult problem still. A hunter must be bred with much substance and also with size. But do not make the mistake of confusing size and substance! As a rule we do not require much over 16 hands in the mare, nor much over 16 hands 1 inch to 16 hands 3 inches in the stallion. The conservative way of breeding a hunter is from a good, well-made, deep, short-legged Thoroughbred stallion, and it is undoubtedly difficult, in a general way, to improve upon that procedure. Yet, under suitable circumstances, equally good results may be obtained from an Arab stallion and in certain cases the result may be even better. The Arab is more potent even than the Thoroughbred in passing quality, and from suitable mares will breed plenty of substance and quite sufficient size.

With regard to cross-breeding, for hunters or ponies or whatever else the purposes may be, I have already pointed out that there must be a measure of affinity between the animals to be crossed. That condition applies not only to the type of animal, but also to size. It is a bad practice to send small mares to stallions that are too big for them. A Thoroughbred stallion of about 16 hands 2 inches should not be chosen to cover any mare much under 15 hands; Arab stallions should not cover anything under 13 hands 2 inches, and ponies below that size ought to be sent to a pony stallion. And as regards the sending of Arab or other pony mares, or even Thoroughbred mares, to a carthorse stallion, that is, in my opinion, as wicked as it is stupid.

Whilst on this subject of breeding hunters, mention must be made of the so-called premium stallions, to whom so many owners send their hunter mares. Premium stallions are on the whole good horses, that are very suitable as to type and quality for the breeding of light horses and particularly of hunters. Some of them are in fact quite outstandingly good and beautiful horses, and they serve at the very cheap fee of £80. Each premium stallion is given a district, except for extensive hunter breeding areas such as East Yorkshire, Carmarthenshire and North Devon when up to 3 premiums are awarded for the same locality. Whereas in the past these stallions travelled along a fixed route, they now stand at a particular place and the mares are brought to visit them. This is considered much more satisfactory for all concerned.

It is a system for the breeding of a reasonable type of stock from indifferent mares. But it is hardly advisable to send the premium stallion a first-class mare, and that for a number of reasons.

Firstly, because as there is mostly only one premium horse in the district, it is not much better than a toss-up as to whether or not that horse suits our particular mare.

Secondly, although these horses only cover on average between 38 and 60 mares in a season they may sometimes cover as many as six in one day. So if your mare happens to be number six on the day, her chance of being got in foal is about nil. The fertility percentage of these premium horses is bound to be lower than usual, normally between 63-66%.

Thirdly, because even if your mare happened to be number one on the day, it would still be a matter of luck if she happened to be sufficiently fully in use to be served with the best chance of being fertilized, if you have not got anything to try her with at home.

Lastly, because these horses get so many bad and indifferent mares that, however good they may be themselves, they are bound to leave a number of bad and indifferent stock behind, and probably, in the course of years, far too many of them. If that gives the horse a poor name as a sire, as well it may do, however undeservedly, that cannot fail to reflect adversely upon the future value of your foal.

But the low fertility is really the gravest risk of all, for if your mare proves barren at the end of her year there will not, if you count the cost of her keep and that of a year of her life wasted, be much left of the saving you thought you could make by paying this low fee.

I must here impress upon my readers that this question of barrenness in mares is in fact the gravest risk with which all breeders are faced. Some mares have under-developed ovaries, or for other reasons either cannot breed at all or are difficult or unlikely breeders. If you have sent your mare to a reliable stud, where people know their business, you are certain to be told, since your mare will be examined, so you will be advised as regards her possibilities and need not go on wasting time and money. Better still, if you can find a good horse vet who is used to examining mares near your home, have her examined before you buy the nomination.

Other mares are shy breeders, or difficult to catch just right, or irregular in their seasons, and some may break the service and yet show so little that none but the most observant stud-groom would ever notice it.

These are all reasons why the chances of success are enhanced so much by suitable environment and by first-class skilled attention.

And even when all these precautions have been taken, we must still allow for disappointments. It is a good average if over 70% of the mares conceive and two out of every three served mares produce living foals, and a credit to any stallion. The cause of the lack of greater fertility is usually with the mares and not with the horse.

It is as well to bear all this in mind so that, once having made your nomination, you can do your bit in helping to achieve the desired result. You can do so by sending your mare in the right condition and at the right time, and by supplying the stallion owner, in writing, with fully comprehensive particulars of her.

Mares have an annual breeding cycle and tend to 'switch off' altogether in the winter months until the hours of daylight start to lengthen again in the spring, and the grass grows. This is known as *anoestrus*. The ovaries are completely inactive with no evidence of follicles or *corpora lutea* and there is a low blood progesterone level. Outwardly the mare appears to be in identical condition to one in *dioestrus*, but in this case the ovaries are active and a *corpus luteum* is present, with a high blood progesterone level. The former will normally respond to the drug Regumate.

In order to obtain an early foal from your mare you will therefore have to mimic the spring conditions and thus awaken her hormone cycle some few weeks before this would occur naturally.

To speed the onset of a normal spring cycle most authorities recommend leaving the light on in a mare's loose-box after dark, long enough to increase the hours of 'daylight' to sixteen hours. Two hundred watt incandescent bulbs must be used for this purpose.

In the absence of any spring grass (which contains naturally occurring oestrogens) dried grass or better still, dried lucerne, should be fed every day.

There is little point in sending your mare to stud until she has started to cycle normally, as until this happens she is unlikely to conceive. Sometimes at the beginning of the year a mare will come in season and remain so for two weeks or longer without putting up a ripe egg. So to save expense, take all necessary swabs but leave your mare at home until she is cycling every 3 weeks.

Mares that are too fat or too hard in condition are difficult to get in foal; they should be in good natural condition, having been turned out sometime previously, if not altogether, at least during the daytime. Horses turned out require the natural grease in their coat, as a protection against wet and cold, and must not be groomed, therefore. None the less, they may be brushed over occasionally and their manes and tails tidied up, and you certainly owe it to her to send her to stud looking creditable and not neglected. But don't send your mare, as happened to me once, in hard hunting condition, clipped out and complete with blankets, rugs and bandages.

The right time depends upon the mare. If she is a maiden, or barren, she should be sent about a week before she is due to come into season; sending her too close to her date may upset her period altogether. If she is in foal, and it is desired that she should be foaled down at the stud, she should be sent a month before her foaling date; travelling too near that date is risky.

If she is to be sent with foal at foot, the fifth day after foaling is the best time to send her; foaling mares are most easily got in foal again during the first period of heat after their foaling, which occurs usually on the ninth day, but may occur at any time between the fifth and the tenth day. Thereafter she will come into season again, subject to irregularities that vary with individual subjects, every twenty-one days, and she may of course be sent to be covered during one of these later periods. Statistics have shown that mares covered on the foaling heat are more likely to prove empty by 42 days after service. Some authorities therefore recommend that mares should not be covered on the foaling heat unless they are foaling late in the season.

As I have said already, mares should be sent brushed over and looking well cared for, with their feet properly trimmed and of course without shoes, at any rate behind. They should travel with a well-fitting head-collar, clean and of good soft leather, preferably with the mare's name marked on a small brass plate. Apart from the fact that this looks well and proper, it is almost essential in helping the stud to recognize your mare and to become familiar with her name; although you may not think so perhaps, it takes quite a bit of doing to sort any given mare out of a bunch of perhaps forty or fifty strange horses. It is most essential that this should be possible, for every mare is an individual and should be treated as such, and in the way that happens to suit her best; it all means a very great deal.

For the same reason you will do well to supply the stud, in writing, with the fullest possible particulars concerning your mare as to name, colour, age, breeding, number of previous foals (if any), name of stallion to whom she is in foal (if that applies), and service dates. Give them also details of vaccinations and full particulars of any idiosyncrasies the mare may have, including temperament and vice, if any, and anything

95

that you may have noted with regard to her seasons or anything else that might affect her from a breeding point of view. Don't imagine that any stud is capable of memorizing all these things in connection with dozens of strange mares, so be sure to put them in writing. Give them at the same time your full postal address, your telephone number, and clear instructions as to how long you want your mare to remain at stud.

The latter is a matter that is entirely up to you providing you keep within the terms of the nomination agreement. For the stud fee that you have agreed to pay you are entitled to one service the first time the mare comes into season, and to as many subsequent services as may be necessary if the mare should break to the first service, until the end of the breeding season. Most studs close their breeding season on July 15th, and although exceptions are often made when they appear necessary or reasonable, you are not entitled to any further services after that date. After all, the thing must come to an end some time, and the stallion will need his rest.

In very many cases one good service, given at the right time, is effective; if the service has been somewhat uncertain, as happens now and again especially with maiden mares, the stud may decide upon a further service. If the mare is still in season 2 days later she should be served again.

But although the majority of mares appear to hold to such single service, some break service and come into season again, at three weeks or at six weeks, and more rarely at nine weeks, from the date of such first service. A good many people are satisfied if their mares hold at six weeks, but it is safer still to wait for their trial at nine weeks, which means leaving them at stud for ten weeks or so to allow for irregularities in the periods. In order to save money on keep fees, some people prefer to take their mares home once they have been scanned in-foal at 17 to 19 days after service. This is fine if they have a good horse veterinary surgeon who specialises in stud work and also some means of trying the mares. Although most mares come in season once every 21 days, some will break anytime from 10 days after their last service date and so could easily be missed without the necessary facilities. Therefore, unless you are sure you can spot your mare when she is in season, leave her at the stud until she has been tested in-foal at 42 days by which time the pregnancy is established; it could be cheaper in the long run. Better still, wait until she has passed 63 days.

The best months for breeding are undoubtedly April, May and June; those are the luscious months of spring, when the mare's inclination to breed is at its peak; it is the most natural time and the one most likely to show results. After June the natural instincts decline again, and certain mares do not come into season again once the summer has set in.

The mare carries her foal 11 months, or 340 days on an average, though considerable variations may occur; the shortest period on record, for a living foal, is 322 days, and the longest 419 days, or a little over 13 months. However, it is reasonably safe to reckon on 11 months, so that a mare served in May will foal down in the following April.

For ordinary breeding purposes there are nothing but disadvantages, and not a single advantage, to be gained from having our foals born before April. The weather during the preceding months is much too cold, and too wet to allow of turning our young foals out; we shall have to keep them and their dams in yards, and pamper them at the expense of much extra food and labour, and even then they will not thrive as well as when turned out with their dams on a sunny day and into a nice bit of spring grass. Later foals, again, do not do so well, because after June much of the goodness goes from the grass and heat and flies become bothersome.

Matters are rather different with racing stock. They take their age from January 1st of the year wherein they are born, and since they may have to race as two-year-old babies and be broken as one-year-old infants, a couple of months' difference in their degree of immaturity may be quite important. It is for that reason that racing mares are covered as much as possible to foal down early.

With these considerations in mind, you will of course make arrangements with the stud regarding the approximate

97

time when you do intend sending your mare. Though not of much importance in the case of barren mares, that may be turned away to grass on their arrival (if they are not placed under lights) it is very important to make definite and clear arrangements in the case of foaling mares, for whom the necessary boxes must be reserved. Boxes are always at a premium at any busy stud during foaling time, and cannot be improvised.

Subsequently, it will still be necessary to advise the stud of the exact travelling arrangements made, and of the time of your mare's arrival. Whether you bring her yourself or by public transport, unaccompanied horses always travel best entirely loose in a box without partitions, without head-collars and without being tied up. Foaling mares and mares with their foals should never travel otherwise. They must be given plenty of straw, of course, a good feed of hay to help them while the time away, and if the journey is a long one, a feed of corn as well: unless they are travelling to the Irish Republic, then no hay, straw or feed is allowed on arrival.

Most studs will advise you of the safe arrival of your mare, if she comes by public transport. They will advise you also of the date or dates whereon your mare has been served. There are studs that make a habit of advising their mare owners at three-weekly intervals to the effect that their mares have passed their trial dates satisfactorily. But that is not really necessary since, unless you hear that your mare has been served again, you will know that she is still holding to her original service.

The manner of rendering accounts varies from one stud to another; some send their accounts out at the end of the breeding season, whereas others make a habit of sending accounts out monthly as this helps the cashflow. It is always understood that accounts are due when rendered.

Whilst your mare is away at stud she is entitled to every reasonable care and attention, but beyond that studs will not accept responsibility for any accidents or disease. But they will, in case of need, call in veterinary assistance or a farrier, and the charges thereby incurred will be added to your bill.

It is always understood that you are welcome to go and visit your mares, and particularly your foals, but again don't do so without making an appointment, don't do it too often, and don't take up too much of the stud-grooms' time; remember, they are undoubtedly very busy!

CHAPTER SEVEN

Stud Animals and Their Management

Horses in the Wild and Semi-Wild State, and Domestic Horses.

IN THE CHAPTERS that follow I will deal with the management of the various classes of stud animals.

We sometimes hear it stated that horses do much better in the wild state, when left to their own devices, than under the careful management that we extend to our domestic animals. Generally, these statements do not refer so much to truly wild horses as to those reared on the extensive ranches of Texas and similar wide expanses of the New World, where the animals live in conditions that certainly come very near to their wild state.

There, the stallion is allowed to roam with his herd of mares; under those conditions there are no covering problems, infertility is almost unknown, and parturition is simplicity itself. Foals grow up and run with the herd; they are weaned by the mares themselves when the appropriate time comes, and grow up strong and healthy young horses.

No doubt that is all very marvellous and is, to a certain extent, solid enough ground for the opinion, sometimes advanced, that our horses too, would do better if left more to their own devices and if given less artificial management.

I believe, most certainly, that there is some truth in that opinion, but, as I have already said, to a certain extent only.

Under the conditions outlined above, herds of horses live in comparatively small numbers on vast tracts of territory where they have God's acres to roam over. They move from place to place, over miles and miles of country, where they can always find, even in the off seasons, ample healthy and

100

untainted food, and fresh and untainted water for their sustenance, and undoubtedly their travels keep them fit and healthy. And that would apply in particular also to the in-foal mares.

Above all, the number of horses per acre, or even per 100 or per 1,000 acres, is infinitesimal, and as a result the land where they graze remains unsoiled, untainted, unhorsesick, and healthy.

In a country like England, where space is restricted, similar conditions do not, and cannot apply. We are forced to maintain a dense, and in many cases a very dense, population of stock on our land. No doubt, having so little land, we do of necessity take much better care of it. We cannot afford to ranch; we must farm. Maybe we farm it very well, grow richer and more abundant grasses, in short, produce far more and far better food per acre. No doubt that is an advantage by which civilized farming scores heavily over ranching, and may be able to claim to be a sounder economical proposition. But against that, I believe it to be a great disadvantage, or at any rate a leading difficulty of civilized farming, that we are obliged to stock out land so heavily. For it cannot be doubted that it is the density of our stock on the land, be it horses or cattle, which is the main cause of our troubles with infectious disease, viruses, horse-sickness, infertility, worm infections, Infectious Bovine Rhinotracheitis in cattle, and similar afflictions.

And notwithstanding the fact that we do produce richer and more sustaining food on our cultivated farms, it is an open question whether we do in fact produce enough variety of it, and whether we do not fail to provide, in the enforced diet that we offer our animals, certain ingredients whereof they stand in need and which they would have no difficulty in obtaining were they able to roam at liberty. It makes one think a bit when one finds, on turning one's animals into the most luscious young pasture, that they may all show a marked preference for the rough and weedy stuff growing underneath the hedgerows.

There still is a great deal that remains to be learned in the

101

matter of metabolism and the science of feeding, and there may well be something in the line of thought that tends to ascribe certain unsoundnesses to dietary deficiencies.

But however that may be, the fact that our domestic animals cannot possibly be kept under conditions approaching those of their wild or semi-wild state, entails that we must make up for the lack of room, of variety of food, and of enforced and strenuous exercise in the pursuit of food, by adequate management.

Nor are those the only reasons.

Our winters are cold and, above all, damp and wet. Our paddocks poach, and freeze in lumps, and their feeding value when reduced to that condition is about nil.

Also, we breed bigger, stronger and faster horses than those born wild or semi-wild; our breeds have been adapted and improved to suit our purposes by the care lavished upon them, and especially by feeding them more. Those are the advantages gained by our system of breeding and rearing, as compared with Nature's own methods. Although they outweigh by far the disadvantages outlined, it is just as well to remember that we cannot have the one without the other. In reality it is just there that good management comes in, since it strives at achieving the maximum advantage from our more or less artificial methods, whilst attempting to reduce the disadvantages thereof to a minimum.

Finally, it must not be forgotten that our stud animals are domestic animals in the true sense, which have been reared under domesticated conditions for generations. We cannot possibly therefore alter the conditions of their upbringing and feeding too drastically without risking grave disappointments.

I hope my readers will bear this in mind, and resist the temptation, should it ever come their way, of following the advice of certain people who think that we can abandon our young stock and other stud animals to the rigours of our English winter unaided. We most certainly cannot do so without impeding their growth and development, and sometimes also their future health.

None the less, I do believe that it is wise and beneficial to follow Nature as closely as our conditions permit, and rather more closely than is the case at some studs. I believe in hardy animals, in plenty of fresh air and in plenty of exercise. But I believe also in plenty of good food, though not in over-feeding, and provided that this be given I have no fear of cold and inclement weather.

Since these remarks apply, in a general way, to the various classes of stud animals to be dealt with hereafter, they appeared to me useful as a kind of brief introduction to the subject.

CHAPTER EIGHT

Stallions

The Stallion—The Way to Handle Him—How to House Him—Fitness, Exercise and Feeding—Trying Mares—Covering—The Young Stallion— Getting Mares in Foal—Trying after Service—Difficult Breeders— Insemination.

SINCE at any stud where a stallion or stallions are kept it is undoubtedly he, or they, who come first in importance, it may be fitting to deal with stallion management first.

Entire horses do not usually commence their stud duties until they are four years old, and often much later than that. At any rate, up till their fourth year they are normally qualified as colts, and as such their management does not call for any very special measures and will be referred to later under the heading of management of young stock.

However, if a colt be destined for the stud, he will be quite man enough to be a stallion at four years old, although he should not at that age be given more than a strictly limited number of mares, and from that time on his management will have to be on the same lines as for any fully-grown stud horse.

The stallion's attributes are virility, strength, courage and intelligence. In these respects he is considerably more formidable than the gelding. Any stallion worth his salt is bound to be very high-couraged, proud and fiery. Any trainer, with experience of working entire horses, will confirm that their interest is keener, and their intelligence brighter, than that of other horses; they appear to take a more real interest in the work they have to do, and to perform it with greater zest, and frequently with unquestionable pride and pleasure.

All this is subject to their being in true sympathy with the man who handles them. In that respect the stallion's sense is very keen and very subtle; he knows instantly whether his man is tuned in to him, understands him and loves him, or is, on the contrary, anxious and maybe inwardly frightened. There is no doubt at all that the stallion is capable of very real affection for his human master, whom he knows very well, whose attention he enjoys, and for whom he will do almost anything without the slightest trouble. There is no doubt either that the stallion does miss his human friendship, and can and does feel miserable and neglected if he has for some reason, to go without it for a while.

All this implies that stallions are really quite easy and, in fact, delightful horses to handle, for the type of man or woman who possesses the necessary affinity with them, and who has the gift of combining true kindness and under-standing with quiet discipline. The good stallion man, like any good man with any other type of horse, must possess the ability to ensure obedience with a mere word or a touch of the hand. In the hands of such men, stallions will retain their true character, their courage, and their docility.

Unfortunately, not all men are like that. There are more than a few who are nervous of stallions and who try to hide their inward fear from others, from themselves, and also from the horse, by a show of bravado, a loud voice and, sometimes, the carrying of a stick or similar weapon of defence. In so doing, they may deceive themselves, they may even deceive others, but they will never deceive the horse. The horse knows, feels and reacts instantly. With his simple reasoning, which is pure logic, he decides that since there is no confidence between this man and himself, there must be something wrong with the man. Since the man does not give his confidence, so neither will the horse. Which is unfortu-nate, because it does lead, sometimes, to rough handling. Now rough handling is bad for any horse, but it is fatal for the stallion who, on account of his high courage and of his pride, will reply in kind, will lose his confidence and, in the end, his temper. Whereas any bad-tempered horse can be disagree-able enough, bad-tempered stallions can become positively dangerous.

105

It is all a question of handling. The more we treat our stallion as we would treat any other horse, the more likely we are to keep him docile and sweet-tempered.

It is the custom on most stud farms to have the stallion-boxes separate and well away from other horses, usually in such a way that the stallion never sees another horse excepting at covering time. Possibly that is the only practical way on a big stud where there always is a coming and going of strange mares. There is no doubt that the stallion knows every animal that belongs on his stud, and provided he is used to them, he may see them all day and every day without getting in the least upset. But he will, and does, go rather frantic at the sight of strangers in his yard. However, where it is feasible — as for instance at my place, where hunters and other animals used for riding, or in the course of being broken, are about — I prefer to keep the stallions just as any other horse is kept, in their loose-boxes with the top doors open, able to look out and about as much as they like.

Whilst we treat them as just one of a party, I am quite sure that they, in their minds, take a different view and overlook in all probability what they consider to be their herd.

My stallions stand next door to geldings, but not, of course, in such a way that they can smell at each other. None the less, the stallion knows very well what horse should be in the next box. And it is better, if at all possible, not to change the arrangement. In that respect, stallions are rather peculiar; they notice every change of routine at once, they resent it, and it upsets them; also they hate being moved from one box to another.

In the same yard I keep my own foaling mares, or mares with foal at foot, or, according to the season, weaned foals and other youngsters, both fillies and colts, of varying ages.

The stallion sees them let out in the morning and brought in at night; he also sees hunters going out or coming home, or may be my one and only hackney being put in harness. Since he knows every animal on the place as members of his tribe, he takes little notice of them, apart from now and then an approving chortle or a soft whinny bestowed mainly upon his foals, of whom he appears to approve. They do not upset or excite him in the least. He displays considerable interest, but no excitement, when our own horsebox unloads our own horses in his yard.

But the moment we receive strange mares, or for that matter any kind of strange horses in his yard, he gets very excited and sometimes even angry, and we are obliged in such cases to close his top door.

None the less, the horse is a gregarious animal, and stallions are no exception to that rule. Where it can be done, therefore, I would always prefer to keep a stallion in a similar companionable manner; they are happier and more contented; their temper benefits by it. Altogether, it is more natural for them, better for their spirits, and therefore better also for their health. *Summa summarum*, if it benefits their health and their spirits, it cannot fail to benefit their fertility.

But wherever we decide to have our stallion-boxes, it is essential that such boxes be free from draughts, though full of fresh air, roomy and safe. Stallion boxes should not be smaller than 18 feet square for horses of thoroughbred size; they may with advantage be larger. Stallions are apt to be quite lively in their boxes, and even to jump and to canter about in them, so that there must be no obstructions on which they can hurt themselves; many enjoy a good roll, and they should always be kept well bedded down.

Possibly the most important, though by no means always the best understood part of a stallion's management, is to get him fit for his work and to keep him so. Now when I say fit, I mean fit — fit and hard. One sees a good many stallions that are kept beautifully, well-groomed, well turned out, well fed and really fat. But fat is not fit! Certainly, we do want plenty of condition on a stallion, particularly at the beginning of a hard season, but that condition must be hard condition, muscular, with plenty of flesh, ribs well covered up, but not fat. Fat condition is soft and flabby, detrimental to the horse's virile power, and to his fertility. Such conditions are unnatural to a serving male animal; they are undoubtedly much too prevalent and have a good deal to do

107

with the low percentage of fertility of certain stallions, expected to deal with as many as forty mares.

The serving of so many mares in one season, serving some of them two or three or even more times, is a tremendous task requiring a really fit horse, the fitter the better.

Keeping a horse fit is a matter of exercise and feeding, and however important feeding undoubtedly is, I am quite certain that exercise is more important still.

There are various ways of exercising a stallion.

The one most usually practised is to have the horse walked out, for a couple of hours, by his stallion-man. Now a couple of hours' walk on a leading-rein, with nothing to carry, is in truth little enough to do for a well-fed horse in the prime of his life. But still, it is better than nothing, particularly if the stallion-man is conscientious and energetic and does in fact give him a two-hours' walk. Perhaps on a broiling hot day, stallion-men are as human as the rest of us. They usually do quite a lot of walking already on a stud farm, fetching in mares to try, and taking them out again; they are often socially inclined and, being much on the road, they know almost everyone they meet; they may find it difficult to avoid stopping for a chat by the way, or for a glass of beer; and, much though these conversations may benefit the friendly atmosphere around the stud, they most definitely do not help to get the horse any fitter.

Another system practised at some studs, is to turn the stallions out in an exercising paddock after being worked on the lunge first. Sometimes one paddock is kept for the purpose, and several stallions are turned out in turn, one at a time, of course. This is not too satisfactory, as they smell where another horse has been; that upsets them and is undesirable, particularly in the breeding season. A much better arrangement is to have a separate paddock for each stallion. On a good many studs the stallion boxes are away from the main yard, and each box has its own paddock attached, an acre or so in extent, where the horse can be turned out for part of the day to stretch his legs, and to nibble a bit of grass. On account of the latter, it is essential to keep so small a paddock scrupulously clean, and the only way to do so is by having all droppings picked up at least once every day. But however, and wherever, serving stallions are turned out, it is essential to have their paddocks well fenced, to a height of 6 to 8 feet, and in such a way that they cannot, when turned out, see or smell any mares, and preferably not other horses either.

It goes without saying that it is not an inexpensive matter to make satisfactory arrangements of this type, although it is true that it does save a good deal of work. However, from the fitness point of view, I do not consider the method very satisfactory, unless the horses are given walking exercise or are lunged as well. Although it is a good thing, no doubt, for the horse to be out in the sun and air, it does not mean much exercise, since the horse will really do little more than stand about; he will not even exercise himself much by grazing, firstly because the grass in such small fields cannot be other than tainted, and secondly because stallions are often too restless to indulge in grazing anyway. However, in the off-season, when the horse has no mares to serve, he may be let down in condition and does not require so much exercise; in that case this method of turning him out for his exercise may be considered sufficient, and will save a good deal of time and work.

The method least used in keeping stallions fit, but the one that I consider by far the best and most effective, is to ride them for a couple of hours each day. A good long walk under saddle keeps a horse extraordinarily fit. There is no harm either in an occasional trot or canter, but undue excitement should be avoided. It is surprising how many people look upon riding a stallion as a feat of bravado almost on a line with riding a lion, and at any rate as something highly dangerous. Nothing could be further from the truth. Truly high-couraged horses are always generous, and though they may be, and usually are, high-spirited, they are never mean. Admittedly, I would not give a stallion to any but an experienced horseman, or woman, to ride, but they, with proper horse-sense, will not have the slightest difficulty.

Highly-couraged horses, and stallions in particular, are no mount for nervous or timid people, but they will go like lambs for a quiet and confident rider.

At my place, stallions go out exercising alone or, preferably, in company with other horses, either geldings or mares, but not of course with mares that are in season. With a little common sense, stallions are perfectly easy to manage in company. Naturally, one does not want to come too close up to other horses, and one certainly must not touch them. A stallion will be perfectly sensible and completely obedient to his rider, but, as far as other horses are concerned, he will stand on his dignity. He will brook no interference, grimaces or similar insults, and will, when provoked, fight instantly. It only needs ordinary tact to avoid such contingencies.

I have found my stallions, when in company with other horses, inclined to be a little jealous, particularly when ridden next to any geldings rather taller than themselves. On the whole, I find it best to ride them in the lead of a string, where they will go contentedly and very quietly indeed. I have found them rather troublesome when ridden out in the company of another serving stallion, whom they know invariably from his voice, and whose presence they appear to resent. Accordingly, I never send two stallions out together. They do not appear to take any exception to the presence of immature entire colts. It all seems natural enough, and in accordance no doubt with their inborn herd instincts.

When worked in this manner, stallions will invariably be very fit; in addition, they will be contented, happy and feel themselves part of things; stallions seem to possess a great deal of self-esteem, and they appreciate the trouble one takes over them, and they display great pleasure in, and satisfaction with, the work they do. I have definitely known them to pine when left too much to their own devices, when they feel bored and unhappy. When used to their regular ride, they love going out and taking their due place at the head of the string.

Under such conditions, they will be found to be much better, quicker and more effective at service; their fertility

110

will be at its highest. Incidentally, they will be more level-headed with their mares, and more even; they will not, as is often the case with under-exercised and over-fed horses, be much too impetuous on one day, and apparently totally uninterested on another.

So much for exercise.

Good exercise must be complemented by good feeding.

Good feeding means the supply of sufficient wholesome and nutritious food, composed of the right elements whereof the animal stands in need to maintain him in prime condition. By prime condition I mean prime breeding condition, and not prime fat-stock condition.

The main food ingredients at our disposal are oats, bran and hay. Oats are, without doubt, the best energizing food for horses; the value of bran is considerable, especially as an adjunct to oats; but bran and oats are both lacking in mineral content and in quality protein. Well-got hay, is rich in minerals and contains appreciable protein; it is an essential food for the horse, and more indispensable than oats as rations must consist of at least 25% roughage. From these three main ingredients we can make up a perfect diet for our hunters, and for colts or stallions not used for service. But such a diet would, as we shall see in a later and more detailed chapter on the science of feeding, lack sufficient protein, carotene and vitamins to maintain the animal's breeding capacities at their peak.

Whilst therefore we may give a perfectly good maintenance ration consisting of oats, bran and hay in the winter, we must supplement this towards the beginning of the breeding season and whilst that is in progress with natural fresh food. Of this there is nothing better than freshly cut young grass, clover or lucerne, left to wither for an hour or two before feeding. Failing that, we may use carrots or mangolds or swedes, whilst as a further alternative dried grass or lucerne is most valuable.

The exact quantities to feed depend very much on the size of the animal, on his condition, and on the amount of work he does. The secret of a good feeder is the possession of an eye for condition and the ability to adjust feeds accordingly. As a general rule, stallions need not be over-corned, and 3-4 kg. of oats per day, mixed with bran and chaff, and 7 kg. of first-quality hay, should be ample.

When no green feed is available, I like to give three corn feeds a day, in about equal quantities, morning, noon and night; if dried grass is available, I will mix a few handfuls of that instead of chaff through each corn feed.

If green feed is available, say from the middle or the end of April onwards, I cut out the midday corn feed altogether and feed greenstuff instead.

Hay is fed an armful in the morning, an armful at midday, and the bulk of the ration at night. Horses are night feeders, and they do best with ample time to dispose of their hay.

Linseed mash is a great help in oiling the digestive system; incidentally, although linseed is high in protein it is a very poor quality and the mucilage it contains is indigestable. It must always be very well boiled before feeding.

Managed in such a manner, stallions will be found to be really fit for their work and good foal-getters.

As far as that is concerned I need no convincing that there is no better way of getting mares covered, and no more effective one, than the natural way, letting the horse run with his mares. I would not hesitate to recommend it, or to apply it myself, if it were reasonably possible to do so. But for reasons already stated, such methods are seldom practicable under our rather artificial conditions, with our restricted space and the need for measures to control the spread of venereal disease. For these reasons and also because the natural way is simplicity itself, I need devote no further space to it, but will instead go rather more fully into the much more usual methods applied on stud farms.

Since Contagious Equine Metritis was first discovered in 1977, veterinary surgeons, stud managers, owners and stud hands have all been made doubly aware of the necessity for hygiene when handling breeding stock, which has led to improved standards on stud farms in the British Isles. A code of practice was drawn up in 1978 and since amended including for the 1987 season. Details may be obtained from your veterinary surgeon.

Briefly, all stallions and brood mares must be swabbed annually. Stallions and their teasers should be subjected to examination and swabbing of the external genital organs by a veterinary surgeon. The examination takes place after the 1st January but before the start of the covering season. A set of swabs are taken on two separate occasions.

Mares are sub-divided into two categories, i.e. high risk and low risk. High risk mares are those arriving from, or covered by stallions in countries other than France or Ireland, or mares which have been in direct contact with C.E.M. in the previous year or who have themselves been infected in past years. All other mares are considered low risk.

The first few mares covered by a first season stallion are normally thoroughly screened for evidence of C.E.M.

'It is an ill wind that blows nobody any good' and this applies to C.E.M. in particular. Personnel are now only too aware of the highly contagious nature of venereal disease and the risks involved when handling the external genitalia of stallions and mares. Disposable rubber gloves are now worn on every occasion — a fresh pair for every mare.

To begin with, it must be realized that the sexual functions of a stallion, who has not been used for service at all for a period of perhaps nine months, go dormant. We must not expect him to be quick and eager at the beginning of his season, and fit and ready to deal suddenly with a number of mares in succession. It is normal, and natural, for the horses to be slow at first, even very slow, and we must give him plenty of time with his first mare or two and not feel impatient or disappointed. His instincts will be re-awakened by and by, and he will quite soon become quick at his job again.

The most serviceable arrangement for trying and covering mares, and the one in use at nearly all studs, is a covering-yard with a box at one end opening out into the yard. The stallion can then be placed in this box, the top door opened,

113

and the mare to be tried led up to him in such a way that the animals can converse on the subject under consideration in safety. Safety, that is, for the stallion and for the man in charge of him, because mares that are not ready to accept the stallion's advances can express their feelings most forcibly with their heels. If no covering-yard is available, more primitive methods will have to be resorted to, but safety for horse and man must always be the first consideration.

Trying mares is a very exacting, exhausting, and often exasperating job for a stallion; it may well take a good deal more out of him than an actual service. When a horse has a list of thirty or forty mares he may, at some studs, be faced with having to try as many as half a dozen mares of a morning, with possibly not one of them coming into service on that day. It is apt to discourage the horse, to tire him, and to make him listless. That is the reason why most studs, with valuable stallions, employ a so-called teaser, which is an entire used for just that particular job alone. Which is, when we come to think of it, rather cruel on the unfortunate horse. It has the disadvantage, moreover, that mares having been made to take an interest in one particular horse are suddenly faced with a different animal for the act of service. With some mares this unexpected substitution of one male for another does not go down at all well, and although they may in the end submit, and are often forced to submit through the use of a twitch and hobbles, I am not at all certain that a service given under such conditions is the most likely one to lead to conception.

The whole thing is rather too unnatural; and though we may not be able to let Nature do it all, under our stud conditions, I yet believe that it is better to try and follow natural ways as nearly as we can. Therefore, though it is not possible to cut out this trying-business altogether, I am strongly in favour of reducing it to a minimum, with great advantage to the stallion, to the mares, and ultimately to the percentage of foals produced; to say nothing about the saving in time and labour.

Instead of bringing in bunches of mares for trial, I like to have the stallion led or ridden past the mares' paddocks daily. It is surprising how well mares know who is about, and how they react to his presence. Mares that are nowhere near coming into season will, apart from a glance in the horse's direction, take little notice of him; others that are coming near their season will come up to the rails and follow him about, and others still that are ready, or about ready, to be covered, will show the unmistakable signs of their inclination. An experienced man will have no difficulty in making a mental note of the mares that had better be brought in to be introduced to his stallion at closer quarters, over the door of the trial-box. In that manner we may discover most of the mares at or near their right time, and reduce the number of fruitless trials to a minimum.

None the less, there always are some particularly shy mares that will not oblige in the manner indicated; with them there is no alternative, after a time, to bringing them in for trial to the horse every other day until they do come into service.

However, mares are finally brought in for trial in order to see whether they are ready, or will soon be ready, to receive the horse's service. When a mare is not in use, or near to coming into use, she will show her disapproval of the suggestions made to her in the most firm and final manner. If that happens, do not, on any account, insist on teasing her any further; it will not do a bit of good, and it will only sour both her and the horse; experienced stallions know very well themselves, and instantly, whether or not they are wasting their time. So do not insist with a mare in this frame of mind, but take her away at once.

On the other hand, if a mare is beginning to show some interest in the horse, though she may still be inclined to squeal, to strike or to kick, though no longer with quite so much determination, it is as well to give her a little more time, since the horse's attentions will then help to bring her on; it will also familiarize her with him more, and make her more disposed, in time, to submit to his service voluntarily, in the natural way.

115

Many stud-grooms believe in the practice of opening their mares prior to service. When a mare is not in use, the neck of the womb, or cervix, is closed hermetically; it opens when she comes in season, and the opening is at its fullest when the mare is by nature prepared for the horse's service, since it is only then that by this arrangement the spermatozoa, released by the stallion, can find their way into the womb, there to come into contact with and to fertilize one of the mare's ova, or eggs. Whilst one is undoubtedly able to feel with the hand that this condition is present, I do not see what useful purpose can be served in this manner, since, if the neck of the womb happens to be closed, no amount of interference, by hand or otherwise, can have the effect of opening it, and of maintaining it in that condition until the moment of service by the horse. At least, that is my opinion. I believe it to be an unnecessary interference with nature, and I am not at all satisfied that it may not, on occasion, do more harm than good; it may spread infection from one mare to the next if satisfactory precautions are not taken. If any legitimate doubt exists about the condition or possible malformation of the breeding organs, a proper veterinary examination by means of the speculum appears preferable to me, and as far as visiting or boarding mares are concerned. Damage to a mare arising from an examination or manipulation carried out by a lay person, could leave the stud open to substantial claims for the damage caused.

Whilst there may often be some difficulty in finding the right time for barren and maiden mares, it is usually easy enough in the case of mares with foal at foot. As a general rule, such mares will be ready to accept the horse on the ninth day after foaling, though some may be earlier, some later, and some others may not come on at all around that time. Their subsequent seasons should occur about every 21 days from then onwards, so that a mare missed around the ninth day may not come in again until the 30th day after foaling. But all mares are liable to be very irregular in their seasons, so that it is no good going blindly by the number of days; they must always be kept under close observation for any signs —

116

and they may at times be very slight — of their coming into season.

Though it is true that mares with foal at foot are, as a rule, an easy proposition to deal with, that rule, again, is not without exceptions. Mares that have had a difficult foaling, that may be torn inside, or mares after giving birth to twins, may be difficult about coming into season, or difficult to get in foal, even though they are in season. Some mares are so preoccupied with the foals as to prevent them from having their first season on the ninth day; some never come into use at all during the time they are suckling a foal, and from them one can only breed every other year, although hormone treatment *or* weaning the foal early sometimes helps. Likewise, mares foaling very late in the year may, for that reason, not come in again that season.

Although these and similar disappointments are bound to occur, it is none the less certain that the bulk of mares will duly come into season and into service. There is a difference; when a mare is in season she may be coming into service, but not before and not until she has come fully into use. It is only then that she will stand of her own free will, in the natural way, and it is only then that she is ready and able to conceive. Services given at any other time than at this precise psychological moment are so much wasted effort, since they will not get a mare in foal. The closer a mare is covered to the moment of ovulation the better.

That is one of the dangers of using hobbles or any similar form of constraint. Now, though I know very well that an experienced stud-groom is not likely to make many mistakes as to the correct time for serving a mare, there is none the less always that risk, where hobbles are used as a matter of course, that, under the pressure of circumstances with so many mares to be served, some of them may be covered a day or two too early.

Hobbles are used as a matter of course at a great many studs; their main intention is protection of the stallion against injury from kicks. Now, though solicitude for the safety of a valuable stallion is understandable enough, I am firmly convinced that such safety can be ensured equally well, and possibly better, with ordinary natural care and precautions. In fact, if care be taken that mares are covered at the right time only, and at no other time, very few indeed will ever attempt to kick at the horse, though there is just the chance that our judgment may have been at fault and that we are trying to cover either a little too early or a little too late. And against that we can take much more simple precautions easily enough, as we shall see presently.

With hobbles there is, first of all, the possibility of forcing a service upon the mare at the wrong time. There is secondly another, and possibly greater, disadvantage still. Anyone who has ever witnessed the hobbling of any flighty, high-couraged and possibly unbroken mare, and the fight, the struggle and the accidents that may ensue, will soon lose his confidence in them. So, since they are unnecessary on a mannered mare and quite useless and even dangerous on a flighty one, why use them at all? I never do, and object to them strongly.

If we lack confidence in ourselves, in our judgment, or in our ability to prevent accidents by a judicious approach to the mare, there is a much simpler safety device that answers well and is fully effective. It consists of a pair of felt kicking-boots, with a thick felt sole, that may be strapped over the mare's hind-feet; they will take the sting out of any kick. Any competent saddler will make them up.

To prepare a mare for service, personnel should wear disposable rubber gloves at this time and during covering. A disposable tail bandage is put on the mare's tail to prevent any of the loose hairs from entering the vagina at the time of covering and perhaps cutting the stallion's penis. The tail bandage must be destroyed after a single use.

The mare's hindquarters are sprayed down with a mild antiseptic, since washing can introduce infection to a bucket which if used at every covering would spread the infection to other mares on the stud.

The mare about to be served should be taken in a snaffle bridle to the door of the trial-box for just long enough to

118

make certain, as far as one can, that she will stand for the horse. She will then be taken without loss of time, but without hurry, to the covering-place, either in the covering-yard, or else to some suitable place outside. It requires a quiet and experienced man to lead her and to place her in position. As a rule, it is just as well to have a further good man in attendance on the mare. His help may be required to steady her, or for some other reason; covering mares is quite a strenuous job and it requires good men.

The stallion is now led up to some little distance of the mare, two or three horse-lengths behind her and a little to her near side so that she can see him, and be aware of his presence. Some mares, especially inexperienced ones, can be a little awkward at this moment and may try to swing around and face him.

It is the job of the man in charge of the mare and of his helper to steady her, and to get her back or maintain her in position. A lot can usually be done by talking to her and patting her on the neck. There seldom is any difficulty with mares that have bred or have been mated before; but there sometimes is, though not by any means always, considerable difficulty with maiden mares. Often they are merely shy, nervous, and a little frightened. In such cases, I try first to have the stallion led round the mare, at some safe distance, that she may see him and get time to settle down and resign herself. Often that works wonders, particularly if the horse be seasoned and sensible, and not some impetuous young fool. But if it does not work, and provided we are quite satisfied that the time is right and that it is nothing but nervousness, from inexperience, that unsettles the mare and makes her difficult or maybe impossible to handle, we have no option but to resort to applying a twitch. But we will only use the twitch temporarily and for just so long as is necessary to keep her in place and to get the horse to mount her; at that time the twitch can be quite safely, and should be, immediately released.

Mares that have their foals at foot must of course be separated from them when they come to be tried or to be covered. It is as well that the mother should not hear the cries of her offspring, which will upset her and distract her attention from the stallion. Unless her box is sufficiently far away, it is as well to leave someone with the foal, who will as a rule be able to pacify it fairly well and to keep it reasonably quiet.

To lead the stallion we use a so-called stallion bridle, which is none other than a bridoon, whereto a leading-rein is fixed by means of a divided chain that clips on to both bridoon rings; the leading-rein should be about 3 to 4 m. long and be made of strong webbing or leather.

The horse is placed in position as already outlined; usually he will begin by curling up his lips and smelling the air which, to him, is full of alluring smells. He will accordingly get ready for his work, sometimes at once, sometimes not quite so quickly, and sometimes he may be rather slow. It all depends on circumstances. However, he must not be led up to the mare until he is quite ready and fully drawn. At that precise moment, but then immediately, since too much fussing about and delaying discourages and maybe disgusts the horse, he is led up to the mare, deliberately and decisively but if possible without too much impetuosity or rushing, as that would frighten a timid mare.

Now, for the sake of safety, the horse was standing behind the mare, but about a metre or so to the nearside of her. He must be led up to her exactly in that position, somewhat away to her nearside and out of the direct line of action of her heels. So that if the mare should attempt to kick, and we can never be absolutely certain, the stud-groom or the stallion-man leading the horse, and the horse himself, are in a safe position. If a mare means to kick, she will do so upon the horse approaching her and not, in my experience, at the last moment when the horse is quite close up and about to jump her. Therefore, if the mare is disobliging and kicks, there is, in this manner, always ample time to stop the horse and to take him away quickly to the nearside and out of the mare's reach, and preferably rather further. For if the stallion be allowed to turn his quarters too close to the mare

120

and she kicks, he may answer, and we certainly do not want any kicking matches. With a view to the possibility of kicking, however remote, it is a good precaution also to hold the mare's head a little high, when she will not be able to kick quite so easily or so hard.

However, in any normal case the horse is led up to the near side of the mare to within a distance of a metre or so, when he will be given his head and left to jump her without further trouble. The impact of the horse upon the mare is considerable, and her tendency, quite naturally, is to move a few steps forward before she has steadied herself, and before she is being steadied also by the very powerful grip of the stallion's front legs upon her flanks. It is not then necessary, as is so often seen, for the men in charge of the mare to exert herculean strength in order to keep her immovably in place or even in order to push her, stallion and all, in a backwards direction. The latter manoeuvre is certainly most uncomfortable for the stallion, and does not help him one little bit; on the contrary, he will manage to cover his mare much more easily if she moves forward a little, so as to allow him to find his own position. That, again, is the natural way.

The only help useful to the stallion, and which is always given, is for the man in his charge to pull the mare's tail to one side and possibly to take hold of the horse's penis in order to guide it into the mare's vagina.

When the horse has finished his service, the mare's front should be moved a step or so towards her near side, which will assist the stallion in dismounting.

Normally, the horse will require no other help. But in cases where a stallion is required to serve a mare rather bigger than himself, or one rather too fat, so that he cannot get a proper grip on her flanks, it is a good thing for the assistants to take a good grip of the horse's front legs, pulling at them in a forward direction, or to provide him with a neck strap to hold onto with his teeth.

Several ideas said to aid conception have now gone out of fashion. One such was that as soon as the horse dismounted, the mare was taken away and led around for 10 or 15 minutes, in order to prevent her straining and perhaps dislodging the semen. The mare will lose quite a considerable amount of the semen, anyway, which is nothing to worry about since the amount of semen produced in one service, contains sufficient sperm to fertilize a thousand mares! As long as a drop of it remains in the mare and finds its way, that will be quite enough for her to conceive. The mare should be returned to her usual surroundings, either to her foal or to her normal companions in her normal paddock. Horses always get upset to a certain extent when we change their surroundings or their companions, and to do so immediately after service is not conducive to conception.

After covering a mare, the stallion should be taken to his box, and his sheath and hindlegs sprayed with a lukewarm mild solution of permanganate of potash or other suitable non-irritating disinfectant. On no account should his penis be dipped into the solution. If more than one stallion is kept, each should have his own spray equipment, bucket etc., all clearly marked with the horse's name. All too many men are inclined to think that a stallion, having served a mare, should be spared his exercise. That is all wrong. Though he can do, of course, with a little rest after covering, he should have his ordinary exercise; it will not hurt him; on the contrary, he will be all the better for it.

Though the serving of a mare, as described above, is by no means a mystery, it is yet a job requiring considerable tact and experience, a sound understanding of horses, nerve and agility. It is not a job for children, or for the old or infirm. It goes without saying that the manners and tractability of the stallion make an enormous difference. It is difficult enough at times with a reliable horse and a fidgety mare. But if, in addition to having to cope with a difficult mare, we have to deal with a blundering fool of a stallion as well, matters may indeed become most awkward, and degenerate into some-thing like a rough-and-tumble, with considerable danger to man and beast.

With a young stallion, therefore, who may be inclined to be too impetuous, it is quite essential to go about it in such a way that he learns that in this job also he must be subject to discipline, and that it is man and not he who calls the tune and sets the pace. It is most inadvisable to try and cover difficult maiden mares with such an inexperienced horse. Older, quiet and experienced brood mares are a better and safer introduction for him.

Since it is our one and only aim, in covering mares, to get them in foal, it is understandable that so many stud-grooms seem over-anxious to secure these results and that they, in order to make doubly certain, give their mares a second service two days after the first covering. Now, taking the long view, I am not in favour of that practice, since in my opinion it will tend to lower rather than to raise the effectiveness of the stallion's services. With anything like forty mares to a horse, which is at any rate already an unnaturally large quantity, mares come at him to be tried and served quite fast and furiously enough to tax him to the utmost, even when giving only one service, plus the unavoidable number of trials, to each mare. And, of course, it is quite certain that a goodly percentage of mares will not hold to their first service, no matter whether they have been covered once or twice consecutively during their first heat. So that he will at any rate have to try and cover a number of these mares again at their next heat, and some of them again during the next heat after that. Now no horse, however virile, can possibly remain reliably effective under the strain of such an undue number of coverings. Therefore, I hold it to be much better, in the interest of the horse, and consequently also and most decidedly in the interest of the mares and of their owners, to go steady with him and to refrain from using him unnecessarily. By preference not more than once every day, though it is impossible always to adhere to that rule; sometimes the horse may have to cover two or even three mares in one day. But that should be an exception and not the rule, and is to be avoided as much as possible. The advice of a good veterinary surgeon should be sought to determine the very best moment for conception; this should reduce the number of ineffective coverings to a minimum.

Incidentally, there is nothing at all to warrant the belief that two services are more likely to be effective than one service; therefore, when a mare has had one good service it is advisable to leave well alone. Only when a service has been bungled is it necessary to resort to a second covering.

However, as we have said, to have the best chance of success service must take place as close as possible to the moment of ovulation. Therefore if a mare is still in season two days later, the service should be repeated.

If a mare has conceived to her service she will not come into season again that year, that is, not unless she slips. There have been cases known of mares coming into season, notwithstanding the fact that they have been well and safely in foal, but that is so exceptional as to require no further consideration. When a mare has conceived, her current period of heat will run its normal course, and does not cease immediately after she has been covered.

When a mare has not conceived to her first service, or slips, she will come into heat again during her next period, usually 21 days later; see diagram on page 125, or she may break at 42 days or even at 63 days, although that happens less frequently. None the less, all mares remaining at stud must be watched carefully for a number of weeks, and require to be brought in at certain times to be tried to the horse, ideally every other day.

The intervals whereat this is done vary a good deal from one stud to another, as stud-grooms hold rather varying views on what is the most reliable system. The method appearing most thorough that I have seen in operation is the trying of every mare every other day, from 10 days after the mare's last service date.

It is needless to say that trials at such frequent intervals entail an enormous amount of work and are, besides, very hard on the stallion. It is at any rate quite certain that when mares believed in foal are tried to the horse, these trials should be made as short as possible so that mares, who obviously mean to refuse the horse, are taken out of his presence immediately.

124

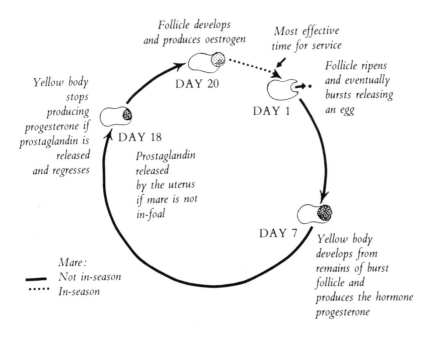

*Follicle develops
and produces oestrogen*

*Most effective
time for service*

*Follicle ripens
and eventually
bursts releasing
an egg*

DAY 20

DAY 1

*Yellow body
stops
producing
progesterone if
prostaglandin is
released
and regresses*

DAY 18

*Prostaglandin
released
by the uterus
if mare is not
in-foal*

DAY 7

*Yellow body
develops from
remains of burst
follicle and
produces the hormone
progesterone*

Mare:
—— *Not in-season*
····· *In-season*

FIGURE 6. SIMPLE DIAGRAM SHOWING THE MARE'S OESTRUS CYCLE

Since mares can be so irregular in their dates, they should be watched carefully at in-between times, and particularly on their 15th, 29th and 31st days; these dates must be noted carefully for each mare, and their behaviour watched with more than ordinary care when the horse comes round on his daily walk by the paddocks. If they do not react to his presence on these occasions, I should leave well alone, but if there be the slightest ground for suspicion that all is not as it should be, they must of course come in and be tried properly.

Mares that hold, and have held, to the stallion for 42 consecutive days from their last service may, as a general rule, be held to be safely in foal. But there never is, nor can there be, any absolute certainty. Mares may not come into

125

service again, notwithstanding the fact that they are not in foal; or, much more likely than that, they may slip their foal, which is so infinitesimal in size during the first few months, that the event usually passes quite unnoticed.

Mares should therefore be examined for pregnancy at 17 to 19 days and again around 42 days. If the owner leaves the mare at stud for a longer period the mare should be examined again at 60 days and 80 to 90 days.

In the later stages of pregnancy, an aborted mare may show the after-birth and make a bag; she may be feverish. In such cases she must be isolated and the vet notified immediately; she should be kept on short rations for a few days in order to dry her up. Abortion is a regrettable occurrence, the more so as it often implies a likelihood of this happening again. If we are aware that a mare has aborted, we must not leave her turned out with other in-foal mares, since it appears that others may be induced to follow her example from nervous sympathy, or more likely from contact with an infection such as virus abortion.

Therefore, when a mare aborts or slips foal she must be isolated immediately and the veterinary surgeon notified. The foetus and membranes should be placed in a clean plastic bag and must not be washed or sprayed with disinfectant first. The bag should be secured and placed well away from other horses, before being taken to a laboratory.

Virus abortion is diagnosed by post-mortem examination of the foetus or foal. If virus abortion is confirmed on your stud notify your breed society and also the Thoroughbred Breeders' Association on (0638) 661321.

However, to return to the normal healthy mare: it is usually safe enough to allow mares that have held for 42 days to their last service, to return home to their owners.

The mares that cause most difficulties at stud are the difficult- and the non-breeders. We may have a mare that keeps on breaking service and comes into season over and over again; others there may be, whose seasons last rather longer than usual, and that appear prepared to accept the horse on any number of occasions; this often happens at the beginning of the year, and also with mares that are infected. It is not much use exhausting the horse and keeping on covering such mares without first examining the possible cause of their abnormal behaviour. One may find a diseased condition, or inflammation, of the mucous membrane of the uterus or of the vagina, accompanied by a discharge 'whose vitiated secretion imperils, if it does not immediately destroy, the life of the spermatazoa, or, should they escape and impregnation take place, the fertilized ovum sooner or later succumbs to its unhealthy environment' (Professor Axe). Such conditions will be revealed by a qualified examination with the speculum, and it is useless to serve such mares until and unless the diseased condition can first be rectified with suitable treatment prescribed by the veterinary surgeon. In fact, to serve these mares would risk infecting the stallion and possibly rendering him infertile and also of spreading the infection to every other mare the stallion covers after the infected mare. All mares *must* therefore be swabbed *before* they are covered for the first time *each* season.

On the other hand, there may be a malformation of the neck of the womb impeding the entry of the spermatazoa in the natural manner; in such cases insemination may have to be resorted to, since with the insemination the seed is injected directly into the neck of the womb. Insemination may also be tried in the case of mares that cannot be induced to accept the stallion's service. However, a check should be made first with the breed Society that the resulting progeny can be registered, when they are the result of artificial insemination.

Then, again, a veterinary examination may disclose weak or cystic ovaries that it may be possible to improve through internal massage or through the administration of hormones. Mares afflicted in this way may not come in season at all until they have been so treated. I have experience of several mares brought into season with great difficulty, and only after

127

repeated injections of hormone, that have bred perfectly satisfactorily afterwards, in two cases that I can remember with only one service.

Brood Mares

ONCE a mare has been served she has become a brood mare. When you receive her back from the stud, with the advice that she passed all her trial dates, or is said to be holding to the horse, she is believed in foal. You, on your part, can only assume that such is the case, and hope for the best. It takes a long time, with certain mares, before they begin to show unmistakable signs of pregnancy. Even then, it may take an experienced eye to decide between a mare that is merely fat and one that is pregnant; and even the most experienced eye may be taken in on certain occasions.

Generally speaking, there will not be much difficulty in deciding, by looking at her, whether or not a mare is in foal at about six months after service; the mare's belly will drop and widen towards the lower part of her flanks; whilst a fat mare makes a nice round belly, the in-foal mare carries her belly more pointed; usually she shows rather larger to one flank than to the other; the best way to view her is from behind and when in movement, or often better still, whilst she is made to take a stride or two backwards; we may then see a slight swinging movement of the load that she is carrying.

However, there are many exceptions. Some mares, particularly mares carrying their first foal, may show very little or nothing at all for a very long time, even, in exceptional cases, right up to the time of foaling. On the other hand, some mares can look so deceptively in foal as to take in even the most experienced.

Although there are a good many cases where there can be no practical doubt of a mare being in foal, one never assumes absolute certainty; for that reason brood mares are sometimes sold 'believed in foal', since 'in foal' constitutes a warranty.

If for any reason we want to make certain whether a mare is or is not in foal, we must have her examined by a veterinary surgeon. There are four possible methods. The first is by internal examination through introduction of the hand and arm into the rectum, in which manner the presence of the foal in the uterus can be ascertained. It goes without saying that this operation is essentially a specialist's job, that must be left to a veterinary surgeon. Even then, I am not fond of this being done, except of course in cases where clauses in the nomination agreement or danger to the mare's health make it essential, as may occur during the later stages of pregnancy. But I am rather inclined to deprecate unnecessary interference of that type with any pregnant animal. For, although it is not supposed to, I have seen it lead to abortion in more cases than one. It does not seem worth while to me to take any risks, however slight, of that description, for no other purpose than that of satisfying one's curiosity.

Moreover, even results of internal examination are not always reliable, as the following example, admittedly rather extreme, will show. A mare that had been served at my stud, a maiden, had gone more than a month over her foaling time. The anxious owner, after having ascertained from me that there was no possibility of a mistake in the service date, decided to have this mare examined for pregnancy by her veterinary surgeon. The examination took place at around 10 a.m. one morning, when the mare was found to be not in foal; on the night of the same day, she gave birth to a filly which, although it was perfectly normal and fully grown, lived for only a few hours.

The second method is known as 'real time' ultrasound scanning. This method was developed in human medicine and later applied to farm animals. It was introduced and has been used extensively on studs since 1981. However, the machines are very expensive so only those veterinary practices which specialise in stud work find it feasible to invest in such equipment.

The mare is *not* X-rayed. The principle relies on the transmission and analysis of high-frequency sound waves. The sound beam penetrates tissues without harming the mare, foetus or operator. The pregnancy can be seen clearly as a dark, well defined, solid circle, in one or other horn of the uterus. It can be detected as early as fourteen days, when it measures some 12-16 mm diameter. By about day twenty-five, the foetus can usually be seen as a small white spot within the black circle. By fifty-five to sixty days after covering, it is generally possible to recognise a tiny head, backbone and legs. Most machines are equipped with a polaroid camera so the developing foetus can be recorded for your interest as well as for insurance purposes. The accuracy of this test lies in the skill of the operator — obviously the more times a veterinary surgeon can use the machine the more proficient he or she will become.

Scanning is particularly useful for the early detection of twins and also for finding mares in prolonged dioestrus — those which have not conceived but have not come in season again.

The third method is the blood test — a blood sample is collected from the jugular vein by a veterinary surgeon between 50 and 90 days after the last service date, and is tested in a laboratory; the optimum time for collection is considered to be 70 days. This test may give false positive results: for instance mares in prolonged dioestrus.

The fourth and last test is the urine test — a sample of urine is collected by the owner any time from 120 days after service and given to the veterinary surgeon for testing. A medicine bottle full is normally sufficient.

From about the eighth month of pregnancy onwards, signs

of life may occasionally be observed inside the mare's belly; at times they may be very pronounced and quite unmistakable, but they cannot be relied on always to make an appearance; in some cases they are never observed at all.

At studs where mares are kept for the special purpose of breeding, such mares are not worked. But they can be worked perfectly well if required. On many farms, it is the usual routine to breed a foal once a year from working mares, and to keep on working them until very near their foaling time. Quiet and steady work does not hurt them and may, on the contrary, do them a lot of good; but they should not be put to any sudden or exceptional strain, such as having to move a heavy load out of a difficult place; neither should they be worked in narrow shafts; chains are safer.

Racing fillies can be rather awkward in the spring, and are sometimes quite difficult and unreliable in their work; they are occasionally put to the horse to get them more settled, and such young mares, freshly in foal, have been known to race well, without any untoward results. Riding mares, too, can quite well be ridden from the first four or five months after service, provided reasonable care be used and undue strain, and particularly excitement, avoided. The well-known show hack June, champion at Windsor and at the White City among numerous other awards in 1946, came to my stud to be served and went on to win the Lady's Hack Championship at Aldershot, remaining quite safely in foal thereafter.

None the less, mares in foal soon tend to become rather sluggish, and there is no great pleasure in continuing to ride them; taking everything into consideration, I would rather abstain from working them once they are in foal, and turn them out to grass to lead the natural life of a brood mare.

Turned out to grass, a brood mare is very little trouble until the winter sets in. But they need company, and good quiet grazing with some shelter from sun and flies. Their company should be quiet and not given to galloping about. Other brood mares are the most suitable, and young stock the most unsuitable. Geldings are completely taboo; they are apt to fuss and tease a mare and may easily, especially in the early stages, cause her to come in season and to slip her foal. Even geldings in the next field, particularly if they can nose and touch the mare over the fence, are equally undesirable. If nothing else is available, a mare will accommodate herself and become quite friendly, with a cow or a goat.

All through the summer and well into the autumn, as long as grazing is fresh and reasonably plentiful, she will require no other food. Good grass is unsurpassed in every respect as the best and most complete food for horses in the natural state. As the autumn advances and night frosts are setting in, the grass will lose much of its nutritional value and will have to be supplemented with a little feed of hay given night and morning. According to what the mare clears up, the quantity of hay given may be increased gradually, until she reaches her full ration of 6.8 kg. a day.

As I get my mares to foal rather late, say from the middle of April to the middle of May, which is the only sensible time for any but a racehorse breeder, I do not begin to feed corn or to bring them in until some time in December, maybe not until Christmas. It all depends on the weather and on the condition of the mares. Generally speaking, they will look extremely well at that time of year, and in first-class breeding condition, neither too fat nor in any way poor.

Breeding condition is most important. Too much fat is not only unnecessary, it is harmful, since it means more difficulty for the mare at foaling time and it also affects the milk supply adversely. However, lack of condition is worse still, since a mare in poor condition cannot possibly do justice to the foal she carries. Weedy foals and neglected yearlings never make great horses; the damage done in those early days, or rather the lack of care taken at that time, can never be made good.

Although the weather in late autumn can be severe enough, it will not do mares in good condition, and which are well fed, any harm; they do not suffer from cold, particularly not from dry cold, but cold, driving rain and soaked fields are uncomfortable for them.

When the weather gets colder and the days shorter, with

133

lack of sunshine, it will be necessary to supplement their ration with corn; it is now necessary to hold them in condition and to see that they do not lose any. If they have a shed or a dry, covered yard to run into, that would suffice to keep them comfortable. It is often said that horses do not use such a shelter, but I have not found that so. Although they are often out on cold, still nights, they will use their shelter and thick dry bed invariably in rough weather.

However, as far as brood mares are concerned, I prefer to bring them into their boxes at night, from about the middle or the end of December onwards. It is easier, in that way, to give individual attention and to make certain that each mare gets her adequate share of feed. They receive an armful of hay first thing in the morning, and 1 kg. of a balanced ration; they are turned out after breakfast and brought in again towards evening, when they are given a further 1 kg. or so of a balanced ration and the bulk of their hay ration for the night; water is always with them. It is not necessary, or even advisable, to feed too much corn; just enough to keep them in nice condition; some mares may need a little more, others less, and some, maybe, can do without it. But on the whole, it is probably best to feed a moderate ration all through.

Boxes should be free from draughts, but airy, fresh and cool. I always have the top doors open whatever the weather, even during hard frost. Fresh air is the best preventative of colds. Besides, if mares are kept in warm and stuffy boxes overnight, to be turned out in the morning into cold and windy fields, we are asking for trouble. And, since exercise is of absolute primary importance for the health of an in-foal mare, it is most essential that she shall be turned out if at all possible. In case of white frost, it is better to wait until it clears up, since the eating of white-frosted grass may cause colic. Otherwise, frost is no detriment; even if the ground be hard and lumpy, the mares can well take care of themselves; neither does snow matter; but the beginning of a thaw, when the top of the ground is slippery on a still-frozen bottom, is treacherous, and under such conditions mares are better kept in. Cold, driving rain is highly unpleasant for the mares if

134

they are to be turned out for a whole day; they will only stand about by the gateway and shiver; in that sort of weather I would rather keep them in altogether, or else turn them out for an hour or two only, while their boxes are being done. It goes without saying that boxes need to be mucked out thoroughly every day, and clean, deep beds provided; it is probable that a warm, soft bed at night does the mares as much good as several pounds of corn.

Whilst it is early enough to bring mares, that are due to foal in April or May, into their winter quarters in December, it is obvious that racing mares due to foal maybe towards the middle of January should come in at least a month earlier. Everything connected with these early foaling dates of bloodstock is very artificial and against Nature, and is, in my opinion, one of the reasons leading to the well-known trouble with low fertility and infertility that causes such great losses to the bloodstock industry.

The régime described above will do right up to foaling time. If our mares foal in April or May they will benefit a great deal by the earliest spring grasses, which are so rich in vitamins and protein. They will encourage milk formation, and will also keep the bowels open. In that respect, it is recommended also, to be rather more liberal with mashes during the last week or two before foaling.

When a mare is to foal down at a public stud and may, therefore, come into contact with disease such as virus abortion, it is a wise precaution to have her vaccinated with Pneumabort-K at 5, 7 and 9 months after her last service date. Some stallion studs insist on this.

Mares carry their foals for a period of 11 months; or 340 days to be exact. At least, that is the normal time, wherefrom a large number of exceptions occur. The shortest time on record for a full-grown normal foal, that I am aware of, is 322 days, and the longest 419 days, as mentioned already in a previous chapter. It is necessary therefore to watch for any foaling signs from about one month before the expected foaling date, which may be ascertained from the gestation table in the appendix. Just about a month before foaling, the

135

bag begins to spring, especially during the night when the mare is at rest; it will go down again during the daytime when she is at exercise in her paddock. However, towards the last week or so, the bag will remain big during the day also, and will assume a shiny appearance. Shortly before foaling, a wax-like substance will usually, but not always, appear on the teats, and this will drop off in from 24 to 48 hours; the mare may then begin to run her milk, which will drop from the teats, and wet the lower part of her hindlegs. More or less at the same time, the muscles round her pelvic bones will soften and drop to both sides of the root of her tail; the vulva will present a swollen appearance.

Under those conditions the mare is likely to foal at any time, and must now be kept under close observation; none the less, she should be turned out in the daytime where she may be seen easily and at frequent intervals. As long as she goes about her business of grazing she may be assumed to be all right, but if she stands in some lonely corner she had better be watched more closely still. All the same, it is but seldom that mares foal in the daytime; they are very shy about it all, and usually manage to put it off until night, when everything is quiet and they believe themselves alone. The most usual time, in my experience, is somewhere between 11 p.m. and 4 o'clock in the morning. When so close to foaling, she must be watched at night. On stud farms there is a special sitting-up-room in close proximity to the foaling-boxes, but any place will do as long as the mare is not being disturbed. The quieter and the more silent one is the better, and the mare should not be made aware that she is being looked at; if she is, it will only make her delay matters more and make her unnecessarily nervous. None the less, if we believe foaling imminent, she may have to be seen about every 15 minutes. If the watchman can hear her, so much the better; again, as long as she can be heard munching her hay, or is standing quietly at rest, nothing is likely to happen for a while. On the other hand, if the mare gets restless, begins to walk round her box, to paw the straw or to get down and maybe up again, it is likely that foaling will start at any moment. So in that case the watchman, if it be that he is not himself the accoucheur, had better go and call the stud-groom; at any rate, whoever is in charge of the foaling can sometimes do with an assistant.

As already pointed out, it is difficult, and in some cases impossible, to be certain of the precise time of foaling. Some mares foal before their time and others go well beyond it. Some mares make hardly any bag at all; others never wax, and others again never run any milk. On the other hand, we may have all the usual signs of readiness for foaling, and yet the mare may hang on to her foal for days and even weeks. Generally speaking, it is not a very good sign if such a thing happens; in my experience if a mare foals much before her time it may well mean twins, and quite likely one of them may be dead already. However, there is nothing that we can or should do to hasten events if a mare is overdue, and we can but wait until Nature is ready to do her work.

I also know quite well that many farmers and other small breeders do not trouble to take any of the precautionary measures set out above, but leave Nature entirely to herself and trust the mare to produce her foal unaided in some corner of a field. Usually that process is entirely successful, since mares will normally foal easily and quickly; in such cases the umbilical cord will rupture, with practically no bleeding, when the mare gets up, and in ninety out of a hundred cases everything will be perfectly all right. Particularly is it so with native pony breeds and the coarser-bred farm-horses. But our thoroughbred mares and their newborn foals are not all that hardy; moreover they are valuable, and above all, there are cases where complications do arise, and when that happens immediate attention is essential. In cases of difficult foaling, due to malpresentation or to other reasons, help, in order to be effective at all, must be procured as promptly as possible. For those reasons I would always have my mares in, and under close supervision.

The foaling-box should be of ample size, and preferably not less than 5 m. square. The greatest danger to any newborn foal is navel- or joint-ill; it is generally supposed that the bacteria causing this dread disease enter the foal's body through the navel string, directly or soon after birth, so scrupulous cleanliness is a fully effective preventative. Incidentally, the mare herself, shortly after foaling, is liable to infections or bloodpoisoning if her bed and her surroundings are not absolutely clean and healthy. I mention these contingencies here and now, in order to stress the absolute necessity of punctilious cleanliness. Any box to be used for foaling should be swilled out from top to bottom, and washed down and scrubbed out with disinfectant, making sure that no corners or grooves, or pockets of any kind, are missed out; the box should be left open to sun and air, and left to dry thoroughly for a couple of days and then be bedded down with good clean straw. If, as happens so often, the mare has to be in occupation for several days before the actual event, greater care than usual must be taken to keep the floor sweet and clean. The bed in a foaling-box should be warm enough, though not too deep, but it is essential to bank the straw well up against the walls and also against the bottom of the door, in order to prevent low-level draughts.

Since this need for cleanliness cannot be over-stressed, I do not wish to leave this subject without stating that the danger of infection to the foal, and in a lesser degree also to the mare, persists for a considerable time after foaling, and is at its greatest during the first nine days. It is therefore not a bit of good to foal the mare down in a nice clean box, and then to shift her to a dirty one a couple of days later on the ground that the foaling-box may be needed for another mare. In fact, the practice, almost universally followed on most stud farms, of having two or three special foaling-boxes wherein dozens of mares are foaled down one after the other, can be a highly dangerous one. For obvious reasons, the action of foaling, with its inevitable loss of blood, after-birth, and so on, is one most likely to leave infectious material about. If, therefore, such a box must be used for consecutive foalings, it is more imperative than ever to disinfect most thoroughly each time. And if, notwithstanding all our care, a case of navel-ill should occur in a box, it should on no account be used again that season. This disease is so highly infectious to foals, and

138

also so liable to become endemic to certain stables and even to entire studs, that precautions to prevent its spread cannot be too severe. In such a case, I would have that box steam-cleaned or soaked in disinfectant, any brickwork rendered in cement, woodwork repainted or creosoted and the floor covered with quicklime, and I would leave that box untouched if possible for several months; the same of course applies to other diseases particularly including metritis and virus abortion.

In addition to cleanliness of the box itself, it will be understood that the person or persons about to attend to the foaling must only touch her, or the foal, with hands that are scrupulously clean or better still wear disposable rubber gloves. For this purpose, a clean bucket filled with a disinfectant dissolved in lukewarm water, a piece of carbolic soap and a couple of clean towels must be held in readiness. In addition, we need a pair of clean sharp scissors, stored in meths or sterilised by boiling, and some wound powder; these articles will be required at foaling time itself. Shortly afterwards we shall, or may, need the following: enema syringe of human size, bottle of liquid paraffin, bottle of castor oil, baby's feeding-bottle with large teat, vaseline, foaling drink, and an ample supply of disinfectant, such as Dettol or Savlon. A normal feed will be kept in readiness. Arrangements will have been made for easy access to the telephone, or for other means of quick communication for the purpose of calling in veterinary assistance if required.

The signs of mares being on the point of foaling vary greatly; textbooks refer a good deal to mares becoming very excited, being very restless in their boxes, and sweating profusely. Such cases do occur, but in my experience they are the exception rather than the rule. As a rule, I have found most mares that have been well looked after, that are in good breeding condition and not too fat, to be quiet and sensible about it. The most usual sign, immediately or very shortly before foaling, is some restlessness; the mare stops feeding and moves about a few times.

The first sign that foaling has actually commenced is the

breaking of the water bag, shown by a flow of yellow/brown fluid (allantoic fluid) from the vulva, which is the end of the first stage of labour. The first stage can take from only 30 minutes to as long as 2 hours or more.

The second stage of labour commences with the appearance of the amnion, containing amniotic fluid. Any stitched mares should be cut open at this point. The second stage which terminates with the complete delivery of the foal, normally takes from 15 to 50 minutes. If either the first or second stages appear longer than normal, or the mare is having any difficulty, it is better to call a veterinary surgeon immediately as the foal may be wrongly presented. For instance, although the fore limbs and head appear to be in the correct position, the hindquarters could be twisted round.

As the mare starts to foal she will probably go down and get up again, or maybe she will stay down and start straining. That implies that foaling is now imminent. It is essential to observe these happenings without disturbing the mare, and without going into her box. The first thing to appear is the amnion protruding from the vulva; usually this will occur when the mare is down, but it may also happen whilst she is standing up; she may even get up and go down with it protruding.

At this stage, a quick check should be made to ensure that the front feet are coming normally. Very occasionally, and particularly in the case of maiden mares, one front foot is pushed upwards through the wall of the vagina into the rectum. Should this occur your veterinary surgeon must be called immediately. However, by careful supervision of the front legs through the vagina this accident can, in many cases, be avoided. If everything appears to be normal, withdraw from her box and watch the mare as before. Within a very short time, usually a couple of minutes or so, the foal's front legs will make an appearance, frogs down, and be visible inside the amnion. Although cases of mares dropping their foals whilst standing do occur, they are, in my experience, very rare indeed. Ninety-nine out of a hundred mares foal lying down. As soon as the foal's forelegs make

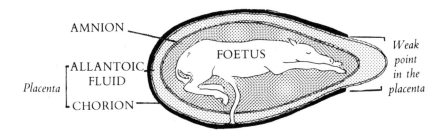

AMNION

ALLANTOIC
FLUID

Placenta

CHORION

FOETUS

*Weak
point
in the
placenta*

FIGURE 7. CORRECT PRESENTATION OF THE FOAL AT BIRTH

their appearance, but not before, it is time to go into her box, and to do so very quietly and without fuss. The person in charge will kneel down behind the mare, and his assistant stand quietly behind him ready to pass him any of the articles he may require. It is most important to keep calm and to do, by preference, nothing at all, except keep an eye on things.

Since the foal has to pass through a very narrow opening, Nature has arranged this in such a way that the foal's shoulders, which are its widest part, come through one at a time. On that account, the front legs should appear one slightly in front of the other; if, as sometimes happens, they protrude exactly on the same level, it is advisable to take hold of one leg and, at the exact moment when the mare strains, but not otherwise, exert a slight pull, not in order to pull the foal forward but only just enough to advance the one leg, and with it the one shoulder, a trifle in front of the other. But, apart from that, we should not on a normal foaling, attempt to take any part in producing the foal. The mare will do everything that is necessary by alternatively straining and

141

resting; when straining, she is exerting enormous muscular power in order to try and drive the foal through the narrow passage; when resting, she not only recovers her strength, but induces at the same time the muscles around the narrow opening of the womb to assume gradually a state of semi-paralysis and reduced resistance. It is, therefore, very wrong to lose one's calmness, and to try hastening matters by pulling at the foal, as is done all too frequently. Very soon after the legs, the nose will become visible, and subsequently the whole head: at this point the bag should be torn and peeled back from the foal's nostrils. So far, everything has been easy, but now the shoulders have to follow, which is much more difficult. The mare will strain vigorously, with alternative periods of rest, and although as a rule even this process does not take more than a few minutes, it may well seem an eternity to the anxious attendant. However, pulling at the foal is dangerous; the risk of forcing the passage, of causing a tear and serious injuries, is great, so that the temptation to pull must be resisted. As a rule, the moment will come when the shoulders are passed successfully, and the foal slides suddenly forward.

The attendant may now support the foal, which will still be in the amnion, underneath its shoulders and, by sliding it gently towards him, help to complete the delivery. As long as the foal is inside this membranous bag and the cord is intact it lives through its umbilical cord, or navel string, and does not breathe; it is important, therefore, to avoid rupturing the cord as long as parturition is incomplete — for the moment the cord is ruptured, the foal will start to breathe, and will from then on have to continue breathing in order to live; now, whilst no danger attaches thereto if foaling proceeds normally and does not take too long, there is a serious risk of the foal suffocating if it continues to be subjected to severe pressure on its ribs once it has started breathing.

However, once delivery is complete, the attendant will at once ensure that the foal's nostrils are clear of mucus and expose the navel cord. Before birth, the foal receives its

142

entire oxygen supply from the blood flowing through the cord from the mare. When the cord breaks, the foal's breathing must be fully established. Premature severance of the cord can result in a considerable loss of blood to the foal. Therefore *all pulsations must have ceased* before the foal is dragged round to the mare's head. As soon as the cord is severed at once apply a liberal dressing of wound powder in such a way that the whole navel is well covered — particularly the end, but also the inside of the navel towards the stomach. Repeat 2 or 3 times daily as necessary.

Now attendant and helper will take hold of the foal, each of them by one pair of its legs, and drag it round to the mare's head, with the foal's back turned towards her mouth. In that way, she can and will lick and fondle it without risk of interfering with its navel. It will help materially to pacify the mare and to keep her quiet.

The attendants will now leave the box and let the mare be in peace and at rest with her foal, until she decides to get up by herself; it is necessary to protect the mare, who will have sweated and be in a state of exhaustion, as well as the newborn foal, from any draught or excess of cold air, and it is most advisable in this case to shut the top door of the box. There is no need to worry about the foal at this stage, since there is no danger of the mother hurting it in getting to her feet. The longer the mare will lie quiet and recuperate the better, but in most cases she is up within half an hour. If the foal be precocious and starts struggling round the box, that is all to the good, and need not worry us; it will help to get his blood circulation and his nervous system going.

Go into the box a minute or so after the mare has got to her feet, and let her have a drink of chilled water and her normal light feed. The after-birth will be seen to be hanging down, and should at once be tied up to the level of her hocks; if hanging too low, there is a risk of the mare or the foal treading on it and so causing its forcible removal, which is always dangerous.

On no account must an attempt be made to remove this after-birth; it must be left to come away by itself, which it

143

will usually do within an hour or two from foaling. If the after-birth does not come away by itself within eight hours from foaling, there will be a definite danger of infection to the mare, and it must then be removed by a veterinary surgeon; it is a delicate and dangerous operation that should not be attempted by any unqualified person. Since, however, eight hours is the limit of safety, I make it a practice to advise the veterinary surgeon at four hours from foaling; it is not a frequent occurrence.

Whilst in the box to see to a feed for the mare, collect and remove the straw that has been wetted and soiled in the process of foaling, and spread some dry bedding over the top; it should only be a matter of a minute or two to pick up the worst of the soiled bedding in a sheet, and no attempt should be made at this stage to remake the bed thoroughly; the mare and foal should not be disturbed more than necessary just at present.

Apart from this short intrusion, the mare and her newborn offspring are best left alone and given a chance to settle down from pain and excitement. It is as well, though, to keep a discreet watch. The foal will soon be struggling to get on its feet, making many abortive efforts, falling over itself, sliding, rolling and struggling round the box, and getting at times entangled in or under its mother's legs. There is no danger. These efforts do the foal a world of good. Finally it will get on its legs and stand there, sometimes within half an hour, but more often within an hour to an hour and a half. It is much better for it to do so unaided than for us to go in and try to help it. It will now find its mother, and its instincts will soon tell it to look for its first meal. It will start sucking everywhere, except in the right place, at the mare's legs, or her hocks, in short, anywhere there is a warm feel. If left to itself, it is ten to one that it will find the proper place in the end, and will drink. The first milk to come from the mare is the colostrum, which the foal needs for its antibody and laxative properties. It is imperative that the foal should have this milk within a couple of hours from birth.

If, notwithstanding all its efforts, it fails to find the teat, it

144

may be advisable to go in and assist it, and to guide its little head to the right place, but do not be in too great a hurry, and give Nature a proper chance first.

Mares are always a little tender in the udder at first, and somewhat nervous and even excitable about this first drink; it is painful to them, and they may well flinch a little and even squeal. But, as a rule, they will not offer any real resistance, and mares that kick at their offspring deliberately are exceedingly rare. None the less, it does occur occasionally with maiden mares that are overstrung, excitable and unduly sensitive. I have heard of mares presenting this same difficulty with every consecutive foal they breed, but have never come across such a case myself. However, when it becomes obvious that the mare will not suffer her foal to drink, and maybe threatens or kicks at it, it is then imperative to go in and lend a hand. The mare should be bridled and held in position against a wall of her box, whilst one attendant holds up a front leg, another guides the foal to the teat and induces it to drink. Use as little force or constraint on the mare as is possible, don't punish her, but try to comfort her and to get her to settle down. More often than not, things will be easier at the second attempt, to be made a quarter of an hour or so later, and as a rule the foal will be accepted presently without any further human interference. But there are cases where anything up to twenty-four, or even forty-eight hours, may pass before the mare can be relied upon to accept the foal willingly; in such cases there is nothing for it, but to keep her under close watch and to go in and make the foal drink periodically until she does in the end accept it.

It may happen that a foal is too weak to get up within the first few hours from foaling and, since it must be fed within the first couple of hours or so, we may have to take a hand. Instead of trying to lift a weak foal on to its legs and holding it there, carrying it more or less under the mare's belly in order to make it drink, which is usually unsuccessful, and an exhausting job for everyone concerned, including the foal, it is best to resort to the feeding-bottle. Draw some milk from the mare's udder and induce the foal to suck through the artificial teat, which, with a little tact, is easy enough as a rule; be careful not to choke the foal. Only in cases where the foal cannot be induced to suck should spoon-feeding be resorted to. In cases of very great weakness, when it is judged necessary to stimulate the heart, a teaspoonful of brandy or, if not available, whisky, may be mixed to about a cupful of milk. As long as the weakness continues, a little feed should be given every hour to start with, reduced later to every two hours.

If we have the great misfortune of losing the mare, the problem of bringing up a newborn foal is not simple, unless another mare can be induced to accept the little waif together with its own. Some nice, quiet and somewhat elderly matrons often oblige very willingly, provided the introduction is made tactfully and the foster-mother held on the first few occasions. I have been able to get a mare to accept, quite obligingly and even with apparent pleasure, the occasional visits of a foal whose own dam produced insufficient milk; in the paddock this foal would leave its own dam, when run dry, and go on a visit to its foster-mother.

Failing a foster-mother, there is nothing for it but to hand-feed with a proprietary mare's milk replacer fed according to the manufacturer's instructions. The milk must be warmed to blood-heat, or about 100 degrees Fahrenheit. However, such troubles are fortunately rare, and we need not fear them overmuch.

Before the foal is born, it is maintained in relatively sterile conditions, but at birth it suddenly comes in contact with every germ in its immediate surroundings. In order to survive, it must develop immunity to these organisms as quickly as possible. Unlike human babies the foal does not receive any resistance to disease before birth, but has to acquire all its immunity from the 'first milk' (colostrum) in the first 24 hours after birth. The level of antibodies in the colostrum can be specially raised against tetanus and influenza by vaccinating the mare one month before she is due to foal. If the mare is not vaccinated against tetanus, then tetanus antiserum should be given to the foal soon after birth and repeated at one month old.

The colostrum contains high levels of antibodies and the ability to absorb these across the small intestine ceases within 36 hours. Antibodies are substances found in the blood which destroy germs that would otherwise cause disease.

After 36 hours the foal can obtain immunity only artificially, by means of injections, or by natural development over a period of time. Where a mare has been running her milk before foaling, the antibody level will probably be low, so an alternative supply of colostrum must be found. This *must* be from a mare or donkey, but not from a cow, ewe or goat, as these do not suffer from most of the equine diseases and most have not been in very close contact with horses to develop the necessary antibodies.

It is a wise precaution to take a little colostrum after foaling from a mare which has not run her milk (as long as she has not produced an haemolytic foal). Build up a colostrum bank in your own or a neighbouring breeder's deep freeze — this will keep for a year, providing that it is kept deep frozen, and is then available in emergencies. The easiest time to milk a mare is when her own foal is sucking. When a mare dies foaling it is often possible to draw a little colostrum from her udder for a very short time after she has died. In the case of a normal healthy foal it will soon have started feeding from its dam in the way that Nature intended for it.

Now, the very first most important and essential thing to watch for, is that the foal passes its first dung; some will do shortly after they have got up and before their first drink, but that is not usual; as a rule, they will only start straining after they have had their first drink. Presumably, under natural conditions, when mares are entirely grass fed, constipation troubles in foals do not occur. But, whatever may be the cause, whether our more artificial feeding or some other reason, the fact remains that a more or less constipated condition in stud-born foals is very prevalent particularly overdue colt foals. In so young an animal, whose resistance to setbacks is only slight, it will soon become serious, and even fatal, if neglected. For that reason I make it a practice, shortly after the foal has had its first drink, to ascertain that there are no hard faeces in the back passage. Have the foal held by an assistant, who passes one arm in front of its breast and the other arm behind its quarters; well oil or vaseline one finger, and introduce it gently and very carefully into the rectum, and remove, again carefully so as not to scratch, any hard dung that may be present, usually in the shape of small and very hard balls of a darkish colour. As there may easily be some similar obstruction higher up and out of reach of the finger, now give an enema of half a pint of warm soapy water, at blood temperature.

This done, everything should be set fair, but afterwards a watch must be kept until the after-birth comes away. This should be checked to ensure that it is intact and no small piece remains inside the mare. This would cause a serious inflammation of the uterus so it should be removed at once, to be buried later, and mother and child left to their own devices for the rest of the night. Although the risk of the mare injuring her foal by stepping on it is so small as to be almost non-existent, it is none the less a worthwhile precaution to leave a light on in the box for the first night, especially in the case of maiden mares, or of any others that appear excitable.

Most of the description given above refers to an ordinary straightforward foaling, when everything goes right. Then things are indeed easy, simple and quick, in fact even more so than may appear from a written description, which it is difficult to condense too much. On the whole, mares foal easily and quickly, but there are exceptions. And the one circumstance that renders foaling such a hazardous affair at all times is that if things do not go right, assistance, in order to be effective, must be procured very quickly. Whilst some animals, like cows for instance, are able to stand up to a difficult parturition for a considerable time, and can wait for the arrival of assistance for a good many hours, this does not apply to mares. With them, when help is needed, it must be provided with the minimum loss of time, or results may well be fatal.

Now the ease of foaling depends in the main on the presentation of the foal. The normal position is the vertebrosacral position, wherein the foal lies in the uterus with its forelegs towards the exit, and its head lying on them, also facing the opening (see fig. 7). That position ensures the normal and easy delivery that has been described in the foregoing pages. In that case, no serious complications are to be feared. Normally, the delivery will be completed within a few minutes from the appearance of the head.

But cases do occur when the mare seems unable to pass the foal's shoulders unaided; the periods of rest between her strainings will lengthen, the strainings themselves become weaker, and the mare is obviously exhausting herself. If that occurs, and no progress has been observed for, say, four or five minutes, it is advisable to help. To that end, let two people kneel down behind the mare, if she is lying, or stand behind her, should she be standing; let each person take hold of a front leg of the foal, above the fetlock joint; as the legs will be very slippery, take hold of them with a Turkish towel in the hand so as to ensure a better grip. Ensure that one leg is slightly in advance of the other: by about the length of the fetlock to the hoof. This discrepancy seems to aid delivery. Now wait until the mare strains again; as soon as she does, and for as long as she does, pull towards you in a downward direction, towards the mare's hocks. Pull with a good deal of sustained energy, but pull evenly and without jerks, and do not attempt to exert brute force; there is no hurry, and a fraction of an inch gained at the time is sufficient. But above all, stop pulling immediately the mare ceases straining, as otherwise you may cause injury and you will certainly do no good. Do not attempt any herculean methods, such as pulling with a number of people at the end of ropes tied round the foal's legs; if you do, you will be likely to ruin your mare. Besides, it is quite unnecessary, as a couple of minutes of combined, gentle and careful effort are bound to do the trick.

Apart from the normal position of the foal, a great many varieties of abnormal positions, or so-called malpresentations, may occur. The least troublesome of these is the lumbo-sacral position, wherein the hindlegs are presented

149

first, frogs upwards, and the head comes last. Generally speaking, this presentation causes no undue difficulty or danger, but it will usually be necessary to assist the mare by pulling, on the hindlegs this time, exactly in the manner described in the preceding paragraph, as should the cord be trapped and the blood supply to the foal interrupted before the head is out, the foal will suffocate.

This book has no pretence at being a veterinary treatise on difficult parturition cases. Besides, if it were, it would be of no value whatever to the ordinary mare owner or stud-groom, for the excellent reason that we cannot possibly become obstetric experts from reading a book. The know-ledge, the touch, the tact and the certainty needed, cannot be acquired without years of study and practice. It is most definitely not an amateur's job to try and deal with a difficult foaling case. There are admittedly certain very experienced and very expert men, with years of stud-groom practice behind them, who will, and can, deal with a good many contingencies that are quite beyond the scope of less experienced people. But, as a general rule, it is an imperative necessity to send for the veterinary surgeon immediately as soon as we discover that things are not proceeding normally. Tell him what he is wanted for, so that he is sure to bring everything required. Have plenty of hot water, clean towels and soap ready for him; he is sure to need them.

A certain sign of serious trouble is the appearance of the water-bag without any further developments; if the water-bag is there, and the mare does not strain or gives up straining, there is grave malpresentation; she is intelligent enough to realize that straining threatens a rupture or a tear, and that she is unable to drive the foal towards and through the opening.

It is to be noted that appearance of the front legs, if not followed by appearance of the foal's nose and head, indicates that the head is facing backwards; in such a case it cannot possibly pass the opening unless it be first moved into the right position. This is essentially a veterinary surgeon's job, and may require pushing the foal back into the uterus before it can be achieved. It is therefore the height of folly for amateurs to attempt any pulling at the front legs, except as described in the preceding pages, when the head has already appeared and it is merely a matter of assisting the foal's shoulders through the passage.

A grave, though fortunately rare, after-foaling accident is an inverted uterus, when the womb is turned inside out and hangs down from the vulva and on to the mare's hocks. It is always dangerous, and if not attended to as promptly as possible by a veterinary surgeon, it will be fatal. While the condition persists, and pending the veterinary surgeon's arrival, it must, in order to avoid congestion and infection, be supported by two men, one on each side, who will support the uterus level with the vulva on a clean blanket soaked in warm mild disinfectant; it is advisable to place a similarly treated blanket over the top, to guard it from cold, and prevent it from drying out.

This accident may be the result of violent after-pains and continued straining of the mare after the foal has already been born.

Such conditions may arise in any case, but particularly if the mare has had a bad foaling. When everything is as it should be, mares are generally quiet enough and will not give evidence of much after-pain or of straining. If they do, and at any rate after a difficult foaling, it is as well to take precautions, and call the veterinary surgeon. Some older mares will, very occasionally, haemorrhage from a uterine artery after foaling.

CHAPTER TEN

Mares and Foals

Laxative Food—Constipation—Diarrhoea—Care of Navel—Navel- or Joint-Ill—Umbilical Hernia—Need for Exercise—First Outing— Teaching to Lead—The Secret of Mastery—Picking up Feet—Turning Out in the Daytime, April to June—Turning Out at Night, June to September—Discipline and Bribery—Weaning Time—Drying Off—Care of Weaned Foals—Worms.

IT IS ADVISABLE to leave the mare and her newborn foal quietly in their box for the first couple of days. If the weather is cold, I like the top doors closed at night and in the early morning; but whenever it is possible to admit a bit of spring sunshine, I have the top doors open.

The box will be kept fresh and scrupulously clean. The mare should be given her normal diet; if there is any fresh spring grass going, we can cut her some and give her an armful once or twice a day, letting it wilt for an hour or two before feeding; grass should not be fed cold and wet. In this way we will keep her bowels soft and open, and encourage the secretion of milk. The condition of the mare tends, through the milk, to reflect itself in the foal, and if her bowels are nice and soft, it is likely that the foal's will be the same.

None the less, with it we must take nothing for granted. It is always necessary to keep a very close watch on the functioning of a foal's bowels, and it is particularly necessary to do so during the first nine days. If it passes its dung easily, in soft little balls of a light yellow colour, all is well and we need not interfere. But if we see the slightest sign of straining, when the foal will keep its tail up and potter around its box rather uncomfortably, it is necessary to take

152

immediate action. In the first place we must, in the manner already described, clear the rectum of any hard and constipated matter that may be within reach of the fingers. If we find such constipated matter, we must administer a laxative. In mild cases, taken in hand early enough, two tablespoonfuls of liquid paraffin may suffice, and should act within two or three hours; if they do not so act, give two tablespoonfuls of castor oil, and keep the foal under close observation; if there are still no results, in reasonable time, more castor oil may be given and an enema administered as well; although we should not give more aperients than necessary, it is as well to know that foals can stand quite a lot of castor oil and will take no harm, in recalcitrant cases, from anything up to a pint of medicine, provided of course that it is not given all at once.

If the foal is constipated, without there being any hard matter within reach of the finger, there may be constipated matter outside our reach. In such cases, I like to give an enema at once, and a dose of castor oil at the same time.

But whatever is done, must be done as soon as the foal's condition has been noted, and it must show results within a reasonable time. If it does not do so, and the foal remains constipated, goes down and shows signs of pain and colic, veterinary assistance should be sought at once. There may be an obstruction, well up in the bowel, which it is sometimes possible to clear.

Constipation in a foal will usually yield quite easily to simple measures taken in good time; but it is none the less a condition to be taken seriously since obstinate cases can very soon prove fatal.

If, as the result of our medicine, or from natural causes, the foal becomes loose, it is necessary to keep him clean with a tepid sponge, and Vaseline in order to avoid irritation and blistering underneath his tail. Looseness, of a bright yellow colour, is nothing to worry about providing the foal's temperature is not above normal. It may quite easily occur when the mare is in season. But diarrhoea is often a sign of infection; so in most cases veterinary assistance should be called in before the foal becomes ill, as young foals have few reserves and if left untreated for too long will die.

The young foal's navel requires careful supervision, to make sure that it looks nice, clean and healthy; the wound should heal and dry up cleanly, and any chance of infection be prevented. It may be necessary to dab carefully with wound powder 2 or 3 times a day to help dry the navel, especially in the case of colt foals.

The bacteria causing navel- or joint-ill are assumed to enter the foal's system through the navel string, directly or soon after birth. Fortunately, this terrible affliction is a rare enough occurrence, and its advent can be prevented as a general rule by adequate precautions of cleanliness and treating the navel stump at birth with wound powder. However, if a foal is to be stricken with this disease, the symptoms usually manifest themselves in the second or third week from birth. The first symptoms are dullness, fever, reluctance to get up, difficulty moving, and presently, marked lameness. The navel will be found to be wet and suppurating. Soft swellings will be, or will soon, become apparent on one or more of the joints, stifle, elbow, hip, hock or fetlock; these swellings contain large accumulations of pus and are exceedingly painful; the disease is a form of bloodpoisoning, and the foal suffers exceedingly. If left untreated the swellings may presently burst. It is possible, by treatment, to cure the disease if veterinary aid is called immediately; otherwise the foal will probably die, or at the very least there will be some permanent damage, anchylosis, and stiffness of one or more of his joints.

Another accident, but much less serious this time, that may occur from the navel, is umbilical hernia. This is a swelling, usually the size of a hen's egg or maybe even bigger, that occurs directly from and underneath the navel; if it does happen, it will usually do so when the foal is two or three months old. The swelling is caused by the protrusion through the navel of part of the bowel. The accident may be due to hereditary predisposition. If not attended to, the presence of this swelling is unsightly, and it will naturally detract from the animal's value; it rarely leads to any serious trouble. Quite often it disappears naturally; otherwise your veterinary surgeon may wish to apply a lamb ring, which is left in place until it drops off. These *must* only be applied by the veterinary surgeon, since if any of the bowel was included the foal would die.

The accidents, illnesses and indispositions above referred to, comparatively rare though they may be, had here to be mentioned mainly in order to stress the necessity of always keeping a watchful eye on the state of health and well-being of a young foal. A thriving foal is bright and lively, and any sign of dullness should cause us at once to double our vigilance.

However, with normal luck, care and attention, there should be few causes for serious worry.

If everything is going well and normally, the need for exercise, for the benefit of both the mare and of the foal, will have to be given attention from the third or fourth day after foaling. If we breed racehorses and have very early foals, maybe in January or February, this presents a real problem; we cannot turn out such young foals in the then prevailing cold or wet weather. In such cases it is almost imperative to have one or two roomy straw-yards, well protected from wind and weather, or else to have a few boxes with roomy straw-runs attached to them.

But if we breed hunters, or any other type of horse not called upon to race in infancy, we can afford to arrange matters in such a way that foals do not arrive before the middle of April or the beginning of May. At that time of year there is no reason why our foal should not go out in the open, provided that the weather is nice and sunny; I like to give my foals their first outing when three or four days old. I wait for the morning sun to be well up, and to have gained some warmth, say around 11 to 11.30 a.m.; I have the mare led out, on a leading-rein, into a nearby small and sheltered paddock, and let her nibble a little grass, on the leading-rein. I do not let her loose on this first day because she might start galloping about, which might be dangerous to a foal so entirely inexperienced, which has not yet found its legs. Whilst the mare is restrained in this way to a little quiet grazing on the spot, the assistant *who has led the foal*, will let the little fellow go. It will begin by just wandering around the mare and her attendants for a little while. Very soon it will react to the feeling of the warm sun on its back and to the softness of the turf underneath its little feet; it will sense liberty and the lure of the wide-open spaces, which will one day be its destiny; it will suddenly jump, plunge and rear in little orgies of athletic extravagance; it will as suddenly stop dead, and stand as still as a mouse, in obvious enchanted wonderment; it will come up to nose and investigate its dam and her human attendants; of those it will evince no fear, provided they are quite still and sensible; then it will dart off again and start galloping about in circles, round and round, and always near its mother; its speed, its agility and its sure-footedness are amazing; it can gallop, turn, twist, buck, rear and plunge with lightning speed, and with an amount of grace, balance and certainty that are nothing short of amazing in a little animal that only a few minutes before had appeared to have hardly any control over its own limbs on the straw bed of its box. Yet out here, in a suitable paddock, it can do all this in perfect safety and without risking any harm; I have never known a foal to hurt itself in these circumstances.

However, on the first day, half an hour of this kind of violent exercise will be quite enough for our little fellow, though the time so spent may seem all too short for its human observers; no matter how many foals we may have introduced into the great outside world in this manner, the spectacle remains always new and wholly entrancing. It is a sight that I would never miss on any account.

Still, after half an hour, mother and infant had better be led back to their box, there to rest thoroughly from this first great excitement; leading, as I have said, in the case of both mother *and* foal.

Yes, I teach my foals to lead from the very first time they

Champion *Anglo-Arabian* stallion 'Basa' at four years old. His ancestors, back to 1798, were predominantly white or grey; the horse himself, from a vast number of mares of all colours, *was* never known to produce anything but greys — a striking example of the pre-potency of the better-bred parent.

'Basa' at *Home*: kept, like an ordinary horse, in the general yard, with his top door *open* able to look out and about just as much as he likes.

Yearlings at grass. They will not need supplementary feeding when turned out on good grass, from spring until autumn.

Yearlings in winter. They can stand to be out all winter, provided they are fed well and liberally with hay and corn, and have access to adequate shelter.

go out; I lead them out, and I lead them in again. I never allow a foal to be let run loose behind its dam; it is a bad and slovenly practice; it may easily lead the foal into danger and so to accidents; besides, it teaches him nothing. It is the easiest thing in the world, and quite an enchanting pastime,. to teach a little foal to lead; if left till later, it becomes a much more difficult matter and the results are hardly ever quite so good.

For the first few times, we place both arms round the little fellow, the left arm round his little breast and the right arm around his quarters; the one arm prevents him from moving forward, and the other from moving backward, whilst our hands, one against his offside shoulder and the other one around his hip, prevent him from getting away from us.

Naturally, he will at first struggle to free himself from this strange embrace, mostly by trying to plunge forward into our left arm. Since the little chap weighs very little, and is not yet tremendously strong, an experienced person will have very little difficulty in controlling him. If we are not quite sure of ourselves, we may enlist the help of a second person to hold him in the same manner, but from the offside.

We begin the first lesson in the box, whilst the mare is held quietly in place. We get hold of the foal, in the manner described, and it will start its inevitable struggle. We remain as passive as possible, restricting our activity to nothing but preventing the little animal from escaping from our arms. Though he will try, and try desperately for a minute or so, to struggle free, he will suddenly and decidedly resign himself and give up the unequal struggle.

It is a basic element in the horse's mental make-up that he will give up any kind of struggle or attempt at resistance from the very moment he realizes that his struggle is unavailing, that he is powerless, and that man is master. And, since the horse never forgets, this first lesson to our foal is of everlasting benefit; it will be all the more useful because, being quite passive in our resistance, we shall neither frighten nor hurt him, and he will gain, with his experience of our undoubted mastery, confidence at the same time.

Having mastered the foal's first and unavoidable struggle, we shall now have the mare led slowly round the box a time or two, and push our little pupil gently round behind her. Thereupon the box door can be opened and our careful and slow procession towards the first outing begun. It needs a sensible person to lead the mare, who will be careful to keep her just in front of, or beside, the foal, and who will regulate the mare's progress in relation to that of the foal and its tutor, which is bound to be somewhat hectic and uncertain for the first time or two. However, proceeding in this way, we shall lead mare and foal a good way into the paddock, and well away from the gate and other solid obstructions, before we give the little chap his first liberty.

Proceeding in this manner for a few days, we shall probably be surprised to see how very quickly our little pupil learns to lead quite easily and quietly. I then replace the left arm round the foal's breast by a stable-rubber, slung round his little neck; that is soft, and less liable to hurt him than leading him from a head-collar. I now lead him, for a week or two, from this stable-rubber round his neck, which I hold in my left hand so that my right hand is free to control and direct his quarters if required. Very soon I shall need to do no more than just hold my right hand lightly on the top of his quarters, and even that will quite soon become unnecessary.

Meanwhile, when he is a week or two old, I get him used to wearing a proper foal's head-collar, which I put on in the morning before he goes out, and take off again when he comes in; with a little tact, there is not the slightest difficulty in getting our small foal to take to this head-collar, to get him halter-broken in other words, perfectly quietly. After he is thoroughly used to this new contraption, I begin leading him from his little head-collar and give up the use of the stable-rubber. In the beginning, I still have two people to lead the party out, one to lead the mare and the other to conduct the foal. But in a very few weeks one person may lead both, leading the mare with the right hand and the foal with the left.

I can hardly over-stress the importance of keeping this leading method up religiously, since it really means that by doing so we have our future horse half broken, and we save ourselves an enormous amount of trouble at some later date.

Whilst on this subject of elementary education, I might as well add that I get my small foals used from the very beginning to having their legs touched and their feet picked up. I make a point of introducing them to the farrier when that good man is on his round to attend to the feet of the mares; although there is nothing that he can, or need, actually do to the feet of young foals, I get him to go in to each of them, to pick up their feet, and to play about for a minute or so, touching them with his hammer and file.

In that way we get mannered little horses from the very beginning; whilst that is always pleasant, it may well, at times, mean a good deal more, and as much sometimes as a matter of life and death. Young horses are liable to meet with accidents, and to injure themselves; such injuries may not be of much consequence if the patient will submit quietly to the necessary treatment, which may entail disinfecting and cleaning wounds, and possibly bandaging them — and as often as not, such wounds will be leg wounds; on the other hand, if we have to deal with a wild little devil that will not allow himself to be touched quietly, we may well find it impossible to treat him properly or at all.

Reverting to the subject of exercise: whilst half an hour is sufficient for the excitement of the first day, there is no reason why mother and child should not be left out for an hour to an hour and a half on the second, and for a couple of hours on the third day. On the second day I have the mare and foal led out in the manner already described, well out towards the middle of their paddock; but this time I give the mare her freedom and allow her to take her offspring about as she likes. For the first three days, though, I keep them in a paddock to themselves for the sake of greater safety.

After that, there is no reason why she should not join up with any other mares and young foals that are already about. When you introduce such a pair of newcomers to an already established party, there is always a manifestation of curiosity on the part of the older members, a certain amount of galloping about, and threatening faces. It does, sometimes look a little disconcerting and even a little frightening. But, whilst it is certainly worth while to keep an eye on the proceedings of introduction while they last, which will not be very long, there is really no need for anxiety; it all goes according to Nature, and it is fascinating to observe how cleverly and carefully the latest arrival contrives to maintain her own position between her offspring and any source of possible danger to it.

We may now gradually lengthen the time for our youngest foals to be out, until, at about two to three weeks old, they may stay out for the best part of the day; provided always, that the weather is propitious.

Young foals should not be left out for any length of time in cold, windy weather; they must on no account be turned out, or left out, in rain; when rain threatens, they should be brought in; if they do get wet in a sudden shower, they must be brought in, dried carefully, and hand-rubbed, particularly over their loins; they should not be turned out too early and before the dew is off the grass, and should be brought in, in the evenings, before the dew begins to fall. Once they are three months old we need no longer be quite so careful and we can take a few more chances with their hardiness. But, even then, I do not believe in having foals out in really rough weather, and particularly not in cold rain.

From about April to the middle or so of June we can have our mares out from morning till evening, weather permitting, and bring them into their boxes at night.

They will come in to a feed of well-crushed oats, and to a good feed of hay for the night. The oats will help the mare in keeping up her milk supply, and in doing well the foal that she has at foot and the new one that in all probability she already carries. If the grazing be reasonably good, the mare will not require many oats, and she may even be able to do quite well without them. Even though, it is preferable to give her some, since it will undoubtedly benefit her stamina.

Incidentally, it is the only way to teach the foal to eat oats, and he will have to know all about that by the time weaning comes to him. If mother and child come in to a filled manger, the foal will very soon begin to play with a few oats and will gradually learn to eat them. None the less, with good grazing, the quantity of oats fed need not be big, and 1.4 kg. daily, or even a little less, may be quite sufficient.

Once the hot weather sets in, from about the middle of June until the end of August or the beginning of September, mares and foals turned out in the daytime will do no good. They dread the hot sunshine, and the flies will make their life a burden. Instead of grazing, or lying about sunbathing as they did in the lovely spring sunshine, they will only stand about under some tree, swishing their tails, shaking their heads, and stamping their feet on the hard, dry ground. It is this latter pastime that is definitely harmful to a young foal's joints, which will become tender and swollen. Besides, the animals will just do no good at all; they will go back in condition.

During the hot weather we should reverse our routine — bring them in in the morning, before the flies become troublesome, and turn them out in the cool of the evening, leaving them out all night. Again, of course, provided the weather is suitable; we do not want them out all night in pouring rain, nor during a thunderstorm. Horses have a habit of congregating underneath a tree during a storm, and there is always the risk of lightning to consider, apart from the risk of a stampede in really frightening weather.

It is a treat, during the hot weather, to see how much our animals enjoy the comfort of their cool and shady boxes; the foals at full length in their soft straw beds, sleeping away the time of day and growing visibly; the mares, contentedly munching their bit of hay; and in the cool of the evening when they go out, the foals, now all on intimate terms with each other and forming a young horses' club of their own, romping about, full of vigour and of the joy of life. The matrons, true to their instinct of night-feeders, get their heads down to the fresh green grass, supping in comfort and

161

at leisure to their hearts' content, and to the great benefit of their offspring. There is no better time to have a walk round one's stock than this time of a summer evening, when one can see one's youngsters at play, and one's matrons enjoying the fullness of mature equine happiness.

The pleasure will be all the greater because horses, treated and educated in the way described, will invariably be friendly and full of confidence; they will come up to one freely, follow one and give many and unmistakable signs of their pleasure at being visited and taken notice of. But I must here underline the word 'educated' just lightly. For, whilst it is so essential to treat one's horses kindly and with confidence, in order to gain their confidence and kindness in return, I must yet warn some perhaps over-enthusiastic novice readers that kindness must not degenerate into silly sentimentality; particularly not with young horses. Whilst we like them to be fully confident, we yet do not want them over-familiar; if we are not careful, well-treated youngsters are rather inclined to become that way, particularly little colts. We should never attempt to play or to romp with them, and when they become rather too forward in their advances, we must maintain respect and discipline, possibly with a stern word only, or else with a slight, but quite determined, correction. And the one thing that we must never do is to feed our horses titbits; I know full well that it is a temptation difficult to resist, but it is a pernicious habit that is sure to spoil our animals, to make them overbearing, spiteful, and sometimes even vicious. They can be just as friendly, and much more genuinely so, without this piece of bribery.

By the end of the summer, when September is on the wane, the flies will have become much less bothersome. It is then no longer necessary to bring the mares and foals in for the whole of the day. Some studs then revert to the practice of bringing them in at night and turning them out again in the daytime. I prefer not to do that. Our foals are by now pretty hardy, and they can quite well stand being out at night for another month or so. Weaning time is approaching, and if we wish to leave our mares out at night and day after weaning, until around Christmas time, it would be hard on them to bring them in at night now, and so to get them rather soft. Therefore, I prefer leaving them out at night and most of the day, but I must bring them in for a couple of hours each day, in order to keep my foals thoroughly used to eating their corn. They will now be able to account for quite a pound of oats by themselves, and I must therefore increase the combined ration to about 1.8 kg. of oats, with a little soya bean meal added. If this system is followed, it is best to bring them in to their feed around noon, and to turn them out shortly after the men's return from their dinner. As the nights are growing colder, and come on much earlier, it is certainly not advisable to keep the foals too long in their warm boxes and then to turn them out into the cold of the evening.

Towards the middle of October most of our foals will be around six months old, or older, and they will by then be quite ready for weaning. None the less, even if they be January foals, there is neither need nor sense in taking them away from their dams any earlier. Provided our foals have been well used to eating their corn, and that they are in good condition, weaning will not cause the slightest worry or setback. On the appointed day, I bring the foals into their boxes, preferably two foals of about the same age and size, and of the same sex, together in one box. In that way they keep each other company and appear to be hardly conscious of the separation from their dams at all. None the less, as a security, I keep the top doors closed for a day or so, but I open them as soon as I can see that they are quiet, and that they have settled down nicely. Even so, I keep them in their boxes and do not turn them out at all for three or four days, as otherwise they might stampede in sudden remembrance of their dams and in frantic search of them.

I put them straightaway on their regular rations, an armful of hay in the morning, some more hay at midday for the few days during which they are confined to their boxes, and their corn feed of well-bruised oats at night, with plenty of fresh water always with them. It is impossible to say exactly how much oats a foal should have; it all depends on the size of the foal, on its condition, and on its appetite. But, as a general rule, it is pretty safe to begin with something like 0.5 kg. or 0.7 kg. a head, which we can increase gradually in accordance with the way in which they grow and progress.

The one thing that is certain, is that we must do our foals really well all through their first winter; it is then that they make more growth, and make it more quickly, than at any other time, and it is then that we may make or mar their constitution for all time. But with a little good, clean corn, a quality protein supplement and some vitamins and minerals as well as with plenty of really good hay, with plenty of exercise, with company, and with a soft, warm bed at night, we are not likely to go far wrong. Provided the master, or the stud-groom, keeps a watchful eye on them always, and sees them do well. We want to see our weaned foals nice and round and solid, and we do not want to see any dull-coated, potbellied little miseries.

At weaning time, the mares must immediately be taken right away, as far as possible and right out of earshot, and out of sight, from their offspring; that is essential for the peace of mind of both parties, and so for the success of our weaning venture. Naturally, the mares will be in milk, and will require drying off. To that end they will have to go on short rations for a few days, either on a very bare pasture, or preferably in some outlying boxes, where they can be kept on nothing but a small ration of hay and not more than about half their usual ration of water. All this will help to dry up the milk secretion. None the less, the udders will have to be seen daily for the first few days. It is not necessary to draw off any milk to ease the bag as that would encourage a further flow of milk and could cause mastitis. If we get trouble with a hard and distended bag, which is very painful to the mare, we should draw off milk and if this contains clots we should seek the advice of our veterinary surgeon.

With ordinary care there should be no serious trouble, and the mares will soon be fit to be turned back on to decent pasture and on to full rations; but it remains advisable to keep them at a safe distance from their foals for yet quite a while.

Meanwhile, in a matter of three or four days, our weaned foals will have become quite resigned to their separate existence, and they should give every sign of their intention to continue thriving. Normally, I wean two foals together in one box, and I keep them in that way right through the winter; normally, too, I let them feed together from one manger, and as a general rule that arrangement works quite well. But it sometimes happens that one of them is inclined to be jealous and a little bossy, and as often as not it will be the smaller of the two that is master. When that happens, it is advisable to rig up a second manger, or simply to place a separate bowl of corn in the opposite corner of the box; in that way there will be no arguments about precedence at the dinner-table.

They will now be ready to go out again in the daytime; as exercise is so very important for them, it is well that they be out from about eight or nine in the morning until around four in the afternoon; they may then come in to their feed of corn, and be given plenty of hay for the night. Frost and snow on the ground will not hurt them, but a white frost will; if these latter conditions prevail, they should be kept in until the white frost has gone off. They should not be turned out, or left out, in cold, heavy rain; in such rough weather they are much better in their boxes, where they will take no harm even if left in for several days on end. But they must be given plenty of fresh air, and top doors should be left open whatever the weather.

We can continue this régime until spring weather comes round, usually, in the South of England, towards the middle of April; choosing a spell of nice mild weather, we may then turn them out for good, and leave them out day and night. If there is by then a sufficiency of fresh grass, foals, other than racing stock, will no longer require any supplementary feeding; but if grass is still scarce, we shall have to go on feeding 0.5 kg. of corn daily.

Racing stock, on the other hand, require some food to get them big enough to break by the end of their yearling stage; with them, as also at times with young hunter stock being got ready for show, it is usual to continue supplementary feeding, combined with sleeping them in, either during the night or day, depending upon the seasons.

All foals have worms; it is essential, therefore, to administer a worming dose as a matter of invariable routine every 4 to 6 weeks and as soon as they have got over the first shock of weaning, say, within a week or so after separation from their dams.

CHAPTER ELEVEN

Young Stock

Yearlings—Fillies and Colts—Castration—Out-feeding—Winter Shelter —Maintaining Condition—Care of Feet—Examining for Age—Leading in Hand—Teaching a Youngster to Stand—Lunging—Boxing—Age for Breaking to Saddle.

OFFICIALLY, foals are considered to have become yearlings on January 1st following the date of their birth. In actual fact, they do not begin to lose their foal characteristics, and to assume those of yearlings, until well into the spring, coinciding nearly enough with the time when they are turned out for the season and left, more or less, to fend for themselves at grass.

It is then advisable to separate fillies and colts; some of the latter may be very forward, and in that case liable to cause some trouble. And at any rate, once horses are out of the foal stage, it is much better to separate the sexes, since even geldings and fillies do not go really well together.

It is the fate of the majority of colts to have to be gelded. Opinions as to the best time for carrying out this operation vary. Some studs have it done shortly before weaning, and whilst the foal is still on the mare; in my opinion, that is rather too early, and a considerable shock to the foal to be followed so shortly afterwards by the further shock of weaning; also, the foal is not at that age properly prepared in a physical manner. It is much more usual to have the colts done when a year old, and I cannot see a great deal against that. However, if we are prepared and able to keep colts and fillies separated anyway, I much prefer to let them go on for another year, or even for two years; I find it benefits their masculine characteristics, and they develop better. But at whatever time we have the operation carried out, it should

167

be done during the cool spring weather, before the flies become active; late autumn and winter are not so good, because sharp cold may also affect the wound adversely. Well done, by a practitioner who understands the job, the operation is quick and simple, and heals up without much trouble; it is almost essential to keep patients turned out, as their moving about will cause the wound to drain and will usually stop the occurrence of swelling, which is an unpleasant and far from painless complication. Given the normal care and attention that the veterinary surgeon will prescribe, no undue difficulties need be feared. None the less, it does cause some considerable shock, and is a bit of a setback.

Yearlings, once turned out, will require little attention., and will not need any supplementary feeding if on good grass from spring till autumn. They are best turned out by themselves; if need be, they can go with two-year-olds, but preferably not with older young stock; if their companions are considerably bigger, they rather tend to be chased about, to become under-dogs, timid, and sometimes spiteful; they will not be so contented and happy, and they will not do so well.

From the middle or late autumn onwards they, as well as the other young stock, two-, three- and even four-year-olds that we may have about, will require supplementary feeding with hay, and later on with corn also. It is never advisable to have too many horses in one paddock, and that applies the more so when we start to feed out. Too many mouths tend to encourage jealousy, chasing and kicking, and the under-dog, and some individual always is, will not get his share. Corn is best fed out in cast-iron feeding-bowls; they are strong and they are heavy, and the animals cannot upset them. It is an excellent practice to use one or two more bowls than there are animals to feed; in that way when chasing occurs, which is quite unavoidable, the victim will find another filled bowl much more easily; it helps to ensure that all get a fair share.

It is true that yearlings and other young stock, provided they are fed well, and liberally, with hay and corn, can stand to be out all winter; none the less, the cold and the exposure night and day does them no good, and is anything but economical of feed. They do a great deal better if provided with adequate shelter. No doubt it would be best to bring them into boxes, if we have the room and the labour; bringing them into boxes has the added advantage of keeping them in the habit of being handled, haltered and led. However, a good many of us cannot, today, afford either the extra room or the extra labour. And almost equally satisfactory results, with practically no labour at all, can be achieved by the provision of adequate field shelters, whereto the stock have free access. I have for my youngsters a spacious straw-yard, with a gate opening into it from their field, which is always kept open, and in the yard is a roomy covered barn, enclosed to three sides and open to the south. All along the long closed-in side there is a hay-rack which is always kept filled and from which they can help themselves freely; corn is fed out to them once a day, at the rate of about 1.4 kg. each, in bowls placed well apart in the open straw-yard. That arrangement is suitable for a party of six to seven youngsters; in real hard weather they are in there day and night, protected from rain, snow and cold winds, and with a dry place to lie on; they do splendidly, and maintain their condition right through the winter.

It is that which counts, maintaining condition right through the winter; December is easy; January is not too bad; but in February and March the winter begins to tell and to show up the difference between good and indifferent management. If we can preserve a bright eye, a glow on their coats, flesh on their ribs and on their quarters, up to the end of March, we can compliment ourselves on the fact that there is not much wrong with our management. It is the maintenance of condition through the winter that does make such a difference to the future constitution, vigour, beauty and value of our horses.

One of the essential factors in the rearing of young horses, but one which is neglected all too frequently, is the care of their feet. These should be seen to by a competent farrier, who really knows his job, once a month regularly. The horn of the hoof of a young horse may grow at the rate of anything up to half an inch per month. This is rather more than the wear to which the feet of horses at grass, or in straw-yards, are subjected; besides, the wear of the hoof is not even — the toe, which is first to touch the ground, and which moreover is rolled or slightly dragged over the ground each time the horse makes a stride, wears more quickly than the heel. Thus, if the feet are not trimmed regularly, and the heels cut down, the horse is liable to become too high in the heel, with the risk of impairing the action of the frog; it may lead to contracted heels. On the other hand, if the toe and sides of the feet are not trimmed they will become too long, will spread, and that will most likely lead to sand-cracks. But the most pernicious risk of neglected feet is that they will assume a faulty shape; the horse will no longer be able to place his feet straight, as he should, but may be forced, by the malformation of his hoof, to turn his toes either in or out. Now, such a faulty position of the foot cannot be maintained for any length of time without affecting joints and tendons also, and forcing them into an irregular position too. Thus, by mere neglect of a horse's feet, we may quite easily bring about twisted joints or crooked legs, which will of course affect the value of the animal seriously.

If such faulty foot positions are discovered in time, and taken in hand early enough, a good farrier may, with luck, be able to cure, or at any rate to mitigate, the damage. Before deciding upon the line of action to take, a good farrier will consider not only the shape and structure of the foot, but also the effect of any malformation that may be present upon the horse's action. He will observe how the horse places each foot on the ground, the apparent slope of the foot, and the relative position of toe and heel; also the slope and position of the pasterns and of the joints; he will be careful also to observe if the horse moves straight, or whether there is any tendency to dish or to go wide, or narrow, in front or behind. Any such defects may be directly attributable to neglected feet, and even when that is not so, and such defects are of a

hereditary or congenital nature, a good farrier can sometimes bring about a marked improvement.

If the horse has come to turn his toe out, which is a serious
defect since it will cause the horse to go narrow, with the risk
of striking himself and even of falling, the outside of the foot
should be shortened and the inside encouraged to grow. If he
turns his toes in, which is much less serious since it will cause
him to go wide, without risk of striking, the inside of the foot
must be shortened and the outside left to grow.

As we know, the age of a horse can be determined from an
inspection of his teeth. Whereas the indications are rather
uncertain, and require at any rate a good deal of experience
in the case of older horses of nine years old or over, they are
quite precise and very easy to verify, in normal cases, with
youngsters.

For ordinary practical purposes it is quite unnecessary to
master all the intricate detail connected with molars and
tusks, together with that of the many irregularities and
exceptions to the rules that may occur. The incisors alone
give quite sufficient indication, and their particulars are easy
enough to memorize: At birth, or a few days after, the foal
has the two central incisors; they will be milk teeth, of
course.

At from 4 to 6 weeks old, the two lateral milk incisors
appear.

At about 9 months old, he will make the two corner milk
incisors; all his front milk teeth are then complete.

At about 2½ years old, the horse begins to change his teeth;
the milk teeth will fall out, and will make room for the
permanent teeth, in a given order.

At 3 years old, the central milk incisors have been replaced
by two permanent teeth; these are so much larger than the
milk teeth that one cannot help recognizing them at a glance;
a three-year-old mouth is unmistakable.

At 4 years old, the horse has four permanent incisors;
the corner teeth only are still milk teeth; again, a four-year-
old mouth is unmistakable.

At 5 years old, the horse has lost all milk incisors, and the

171

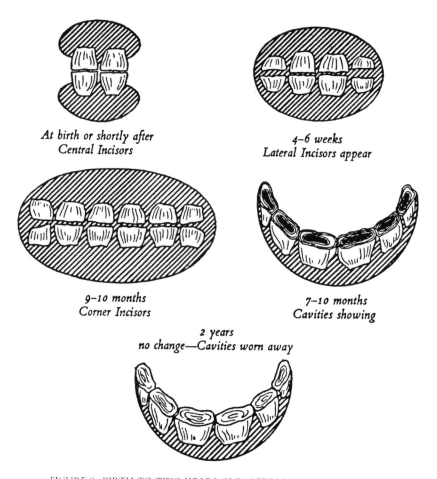

At birth or shortly after
Central Incisors

4–6 weeks
Lateral Incisors appear

9–10 months
Corner Incisors

7–10 months
Cavities showing

2 years
no change—Cavities worn away

FIGURE 8. BIRTH TO TWO YEARS OLD. APPEARANCE OF MILK TEETH

corner teeth are now permanent teeth also; but they are still smaller than their neighbours and they do not yet come into friction. None the less, the horse now has a full mouth.

At 6 years old, the corner incisors have come into full friction.

From the age of six onwards, we are guided by the amount of wear shown by the friction surface of the incisors. As we have seen, the horse enters into possession of his permanent teeth, two at a time, first the central incisors, a year later the lateral incisors, and still another year later the corner

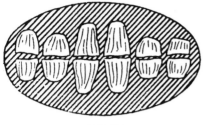
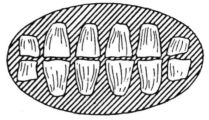

3 years
Central Incisors Permanent

4 years
Central and Lateral Incisors Permanent

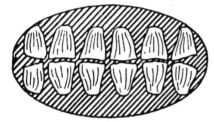

5 years—All Teeth Permanent

FIGURE 9. 3-5 YEARS OLD. APPEARANCE OF PERMANENT TEETH

incisors. It is obvious, therefore, that these teeth come into use and into friction, and begin to wear, at intervals of one year each time. This wear shows itself through the gradual disappearance of the cavities that are present in each tooth when it begins its life-cycle. The cavity shows itself as a black mark, surrounded by an ivory ring; it is very clearly defined in a young tooth. Bearing this in mind, and remembering the order wherein the incisors come through — central first, laterals second, and corner teeth last — it is then easy enough to check up on a horse's age, up till 9 years old.

At 6 years old, whilst the corner incisors have come into friction, the cavities of the central incisors have begun to wear away.

At 7 years old, the cavities of the central incisors have gone, and those of the laterals show wear.

At 8 years old, the cavities of central and lateral incisors have worn away, and those of the corner incisors show wear.

6 years	7 years
Corner Incisors in full friction. All Cavities showing; those in Central Incisors begin to wear away	*Cavities in Central Incisors have disappeared; those in Lateral Incisors begin to wear away*

8 years	9 years
Cavities in Central and Lateral Incisors have disappeared. Those in Corner Incisors begin to wear away	*All Cavities have disappeared*

FIGURE 10. 6-9 YEARS OLD. DISAPPEARANCE OF CAVITIES

At 9 years old, the cavities of all incisors, including the corner ones, have gone. From now on the horse is considered 'aged', because it is no longer possible to determine his age with a degree of certainty from the markings of his teeth.

Young horses that have been treated and educated as foals in the manner described in the previous chapter, will not be much trouble to break when the time comes. None the less, if time allows it, or if perhaps we wish to show our youngsters, it is as well to push their education just a little further along.

The first essential for a young horse is to lead well. To lead well means that the youngster must walk out freely beside, or rather slightly in front of one, on a loose rein, and that presently he will trot out in the same free, brisk, and easy manner. There is nothing that looks more inefficient than a colt that keeps on hanging back and that has to be dragged along; incidentally, in an animal presently to become a riding horse, free forward movement and impulsion are all-important; and so are ease and lightness of control.

Nothing is easier, or quicker, than to teach a youngster just that; it is a matter of no more than ten or fifteen minutes; but nothing is more important in my opinion, than to teach the youngster entirely by oneself, without using any force, and without any outside assistance. It is only so that he will associate the presence of one very gentle, but calmly determined, person, with the need to do as that one person tells him; and it is only thus that he will learn to obey gracefully, without constraint, and with complete confidence.

To that end I lead the youngster out on a leading-rein, slipped through nothing more severe than a head-collar; I walk beside him, holding him lightly on an entirely loose rein. I want him to walk out fast and freely, with his shoulder level with my shoulder. I talk to him and encourage him with my voice. It may not work right away, he may hang back, sulk, or possibly stop. I do not pull at him, I also stop. I carry a light whip in my right hand and with that I give him a light tap, just over the top of his quarters, so that the impact, light as it is, lands well behind his hip-bones. There is no pain, no fright and no shouting, and the animal's confidence is not disturbed. Yet the effect is immediate, and he will move forward at once. It is merely a matter of a little tact and understanding.

From that one step forward, to walking briskly beside one — always using the same aid of a slight tap on the quarters, and the aid will not have to be used very often — is, as I have said, no more than a matter of ten to fifteen minutes. But it is quite enough progress for the first day, so do not ask for more.

Repeat for a couple of days, just ten or fifteen minutes each time. Then ask for a trot — it will cause no difficulty at all; repeat that also for a few days. Now teach him to rein back; don't push or pull; just stand in front of him, say 'Go back', whilst tapping him lightly on his breast with your whip; he will understand in less time than it has taken me to write this down.

As he knows how to go forward, and back, at your slightest indication, it is now quite easy to teach him to stand properly, as a horse should, in front of a judge or of a buyer. Anyone looking at a horse should see him stand on four legs, and not on two as is so often the case. Anyone who knows how to look at a horse, will do so from the near side. Therefore he should stand with the near-fore just a couple of inches in front of the off-fore, and with the near-hind just a couple of inches behind the off-hind. In that way, you show that your horse stands over a lot of ground; yet you do not stretch him, which is only done with hackneys and with carthorses but never with a riding horse. A horse that stands stretched shows hollow and weak in the back, and that would never do for a riding horse. But standing him as explained, you will show your riding horse off to best advantage. The near-fore a little forward will accentuate the slope of his shoulder; his back will be straight; and the near-hind just a little to the rear will show up the length from hip to hock. These are but little points, but they can and do make a lot of difference, and incidentally they show whether the attendant does, or does not, know his job.

Having gone this far with the preliminary education of our youngster, it is as well to teach him, at the same time, how to go quietly on the lunging rein. That again is quite an easy matter, that one can and should perform single-handed in a matter of minutes. People who say, and write, and practise, that to teach a young horse lunging they require two fifteen-stone men in the centre of the ring, and one record fast runner with a whip behind the unfortunate animal, may be excellent horsemasters in many ways, but they certainly do not possess the beginning of an idea about how to break and school a horse.

Once our youngster has been taught to walk out freely beside one, one needs to do no more than to walk him on a circle, gradually getting a little further away from him, and

176

finally standing quite still in the centre of the circle. And it can all be done with an ordinary head-collar, a lunging-rein attached to it underneath the chin, and the aid of the light whip with which the animal is familiar and of which he has no fear.

Finally, I consider it most important, in these days, to teach our young horse to box. A horsebox or a trailer is something of which any horse has some instinctive distrust, or at any rate he lacks confidence in the contraption. There is first of all the ramp, which gives to his weight and which is therefore not solid earth, and consequently something to be rather careful about; then there is the low roof, which impresses the colt, particularly if he is tall, with the fear that he cannot get in without hurting his head. Now, when a horse lacks confidence, or is in actual fear, it is quite essential to eradicate all that and to inspire confidence before attempting obedience.

Therefore the morning of our departure for a show is definitely not a propitious occasion for the purpose of teaching a youngster to box. Shows are usually far away, and young hunter classes are expected in the ring at quite an unearthly hour. Considering that we have to get our youngsters ready first, including the laborious process of plaiting their manes which takes a good man half an hour per horse, that we must allow time for boxing, for travelling, for unloading, for getting them to the collecting-ring, and the one-hundred-and-one other little things that need doing, there will obviously be no time to spare — even though we have, no doubt, got up before the crack of dawn. And even though there were time to spare, our state of nerves on the morning before a show is not likely to be conducive to the absolutely unruffled patience that is required for success with a nervous or frightened young horse. It is hardly necessary to add that the atmosphere on the showground, when trying to box for home, with the inevitable hundred-or-so people crowding round our box to offer their well-meant but quite useless advice, is much worse still. So we owe it to our horses to teach them everything there is to know about boxing, well before the need to put this knowledge into practice arises.

Again, it is not very difficult, though perhaps rather more so than the other lessons already described; this time, since the horse is in some fear, we must on no account use our whip, not even gently, and no force, pulling or constraint of any kind; if we do, the animal will only associate this unpleasantness with the intended process of boxing, and his instinctive feeling of distrust will be accentuated all the more.

I place my box in the yard, with the loading-ramp down, and in such a way that plenty of light falls into the interior so that the animal can see where he is required to go. All partitions have been removed. I support the loading-ramp as carefully as possible, so as to make it feel as solid as may be. The youngster is now led up to the ramp on a loosely-held leading-rein, fixed to an ordinary head-collar. He will no doubt stop in front of it; we pat him, and give him a handful of corn from the bowl that we carry in our free hand. We give him plenty of time; maybe, after a minute or two, the lure of a little more corn will induce him to place one, or perhaps even two, front feet on the edge of the ramp. Maybe he will first want to investigate the ramp more closely, and to have a good smell at it; naturally, we let him do so, remaining ourselves passively encouraging; that implies that we must never, at any time, hold the leading-rein, tight; horses have a definite touch of claustrophobia, and hate being held tight in the face of anything they are afraid of; it will invariably cause them to struggle and, in order to free themselves from the restraint, to rush back; which we want to avoid. If we cannot induce the animal to place his front feet on the ramp of his own free will, we need only pick up his feet, one at a time, and place them on the ramp ourselves. Once we have him there, success is assured, provided we remain calm, passive, and that we are in no hurry. I may take anything from five to twenty minutes, with handfuls of corn, encouraging talk, and patting, but always without using any force, to induce him to mount the ramp gradually, and finally to make him walk into the box.

That is the moment to make a lot of fuss of the youngster; stay with him in the box for a matter of five or ten minutes, and give him a nice feed of corn meanwhile. Then lead him out quietly and in again, two or three times running, giving him a little corn each time inside the box. In this way, in a matter of twenty to thirty minutes, we have gained his confidence and, provided we continue the lessons for a few days, which will now require no more than five or ten minutes daily, we have an animal that will never in his life be any trouble to box.

I like to leave the box, arranged as described, in the yard for a week or so, and to have my youngsters led into it once daily and given a feed of corn.

In my opinion, the above constitutes all that a youngster need know before he is taken in hand to be broken to saddle.

As we know, racehorses are broken to ride at a very early age. That may be in the best interests of the racing industry, but it is most certainly not in the best interest of the young horse. If we have hunter or other riding stock, we need not be in such a hurry. Most hunter breeders break their youngsters to saddle, but only just enough to get them to carry a man quietly, at three years old; usually they are then turned away and taken up again for further education when four years old. Carried out in this way, it is quite a good system of education. With my own horses, I rather prefer to wait another year. As a rule, I break them as four-year-olds, and get them to ride nicely and quietly, mostly at the walk, with a little trotting and cantering later on. But I do no strenuous work with them. At five years old, I take them into regular hacking exercise, I may send them to a few shows, and I take them cub hunting; but I still give them no real hard work, no long days, no big jumping, and no heavy going. Then, at six, they are ready for anything.

I quite realize that many people will consider this system rather slow; as a matter of fact, it is slow, but it does produce sound horses that will last a lifetime. A young horse's bone is brittle and has not finished growing; his tendons are easily injured. We may quite well be lucky with a four-year-old in hard work, but the chances of unsoundness occurring in such

179

a youngster are at least ten times greater than in the mature horse. And, since I make my horses for my own pleasure, and wish them to last, I follow the method indicated; over a period of years I have found it eminently satisfactory.

CHAPTER TWELVE

Feeding

Water—Grass—Constituents of Food: Protein, Starch, Fats, Fibre and Minerals—Vitamins—Protein Deficiency—Lucerne—Maize—Various Kinds of Hay—Oats—Chaff—Bran—Linseed—Beans and Peas— Boiled or Steamed Oats—Feed-barrow—Carrots and Other Roots—Cod- liver Oil—Worms—The Art of a Good Feeder—The Master's Eye.

HORSES require water and food. I place water first, because it is the most essential of the two for several reasons. The horse's body consists, of 70% — young horses 80% — of water. Without constant and adequate maintenance of the necessary supply of liquid, the chemical processes required for the maintenance of health, and of life itself, cannot go on; blood circulation becomes impaired and digestive processes impossible; whereas life can go on for a considerable time without food, it cannot continue for any length of time without water.

The water to be given to a horse should be fresh, clean and untainted. A small quantity of water, such as a bucketful, or even the contents of a much larger vessel, goes off quickly and loses its freshness; it becomes tainted easily through smells and impurities in the air which must always be present in a stable. We know ourselves that a glass of water which has stood on our night-table all night is no longer a good drink in the morning. The same applies to the water in the horse's bucket. It should not be left with him for more than a few hours; the bucket requires emptying, swilling out, and refilling with fresh running water from a tap three or four times a day. In this respect, I am much in favour of the automatic drinking-bowl, which enables the horse to help himself to fresh running water whenever he wants it. The only disadvantage of these drinking-bowls is that they must be cleaned meticulously at least once a day, and unfortunately that is often neglected. When the horse drinks from these bowls he presses a lever, which opens the automatic tap or valve, with his muzzle; in order to make it easy for the horse to do so, the lever in question is shaped more or less in the form of a large sugar spoon; now there always is a little water in the bottom of the bowl, underneath the sugar spoon, which the horse cannot reach, and which therefore remains and goes stagnant. Also, each time the horse drinks, some of his saliva will mix with this water, dust will blow into it during the day, and it will become dirty and contaminated. This explains the necessity of careful and regular cleaning. The alternative is a small automatic water trough. These are normally fitted into the far corner of a long manger and are more suitable for foals since they work on the ball valve principle and so re-fill automatically. However, since many horses like to eat and drink at the same time, they must be cleaned out daily. Most are fitted with a plug to facilitate this task.

There is soft water and hard water. Distilled water is the softest, because the specific salt and minerals, which are in suspension in hard water, are removed by the process of distillation; other soft water is rainwater, and most river water. On the other hand, spring- and well-water are usually hard, and it is considered that quite 80 per cent. of the world's water supply is hard.

Hardness of water is mainly due to the presence of carbonate of lime, and also magnesium carbonate, which are both soluble in water that contains carbonic acid. There are traces of common salt. It is recognized that moderately hard water is more agreeable to the taste, and more refreshing, than soft water, which has rather a flat taste; on that account it is preferable to use it for horses also.

The fact that I advocate the use of automatic drinking-bowls implies that I am in favour of leaving a supply of fresh water always with the horse, so that he may help himself to what water he wants, when he wants it. In that respect I am supported, to the best of my knowledge, by unanimous veterinary opinion. Captain Hayes (*Stable Management and Exercise*) points out that 'as water is the natural drink for horses, it possesses no properties that would induce these animals, under natural and usual conditions, to take more water than the quantity demanded for the requirements of the system'. I mention the matter because the idea that water must be given only three or four times a day, at set times before feeding, is still prevalent in certain stables; to my certain knowledge, that fallacy had been widely recognized as old-fashioned at least fifty years ago. The horse has a small stomach, and should not take any large quantities of either food or drink at any one time; his nature is to eat, and to drink, small quantities often. When water is left with him, he will not overgorge himself; when, on the other hand, his bucket is taken away, he will often suffer from thirst and will be encouraged to drink far too large a quantity of water at once; and even then he will still be suffering from the discomfort of thirst, which is detrimental to health, at in-between times. I have practised the system of always having water with my horses, and I have seen it practised almost universally, for half a century, and I have never yet seen or heard of any untoward result.

The staple food of animals on a stud farm is grass and products directly derived from grass, mainly hay. Grass, at its best, is a complete food, which comprises all the elements which the animal needs for maintenance. One of the major difficulties in its management arises from the fact that one has to deal with a mixed crop of grasses, clovers and herbs. Well managed, it can provide the complete maintenance ration for barren and maiden mares from approximately mid-April onwards.

Good quality hay is the key to feeding stud animals. Any deficiency in the animals' requirements must be made good with other foodstuffs. Explained briefly, the value of a given food depends upon its constituents as to protein, carbo-hydrate, fibre, fat, vitamins and minerals.

Young foals, and mares in the last third of their gestation

183

and during lactation, have a critical requirement for protein quality. The quality of protein in a horse's diet is related to the amino acids it contains. Amino acids are the 'building blocks' from which proteins are made up. About 25 of these have been identified as chemical compounds. The number present in a particular protein varies and farm feeds generally contain more than one type of protein. A mixture of foods is therefore likely to contain a wide range of amino acids, of which lysine and methionine are usually in the shortest supply. In practice diets should contain a modest safety margin of protein to ensure that there is no dietary burden on the mares during stress conditions such as lactation and winter weather; this safety margin also helps to ensure that there is an adequate level of amino acids in short supply.

Good hay alone, fed ad lib, will, in theory, provide the maintenance requirement of barren and maiden mares. However, in practice some concentrates are always fed. This makes good any deficiency in the hay as well as slight differences in nutrient requirements between individual mares. It also keeps the mares in a rising bodily condition before and during covering. The protein requirement of barren and maiden mares stated as a percentage of the total diet, is said to be 8.5% crude protein; that of mares in the last ninety days of pregnancy increases to 11%; but during the first half of lactation the requirement is 14%. Foals require 18% crude protein best supplied as nuts or cubes in a creep feeder.

Protein supplements most commonly used in stud rations to rectify the amino acid deficiency of oats and other cereals include:

Full-fat or extracted soyabean meal; skimmed milk powder or high-grade white fish meal.

Energy producing substances including carbohydrates and fats. Sugar, starch and cellulose are all examples of carbohydrates. Carbohydrates form the largest part of a horse's diet; they are energy and heat producing substances pure and simple; they cannot replace or fulfil the functions of proteins. However, there must be a proper balance between

184

the two. If the calorie to amino acid ratio is correct it maximizes energy utilisation. As the energy content of the ration is increased, there is an increased need for amino acids. A lack of protein in the ration, therefore decreases energy utilisation and increases the cost of keeping horses in optimum condition. The exact energy to protein ratio for the horse is, as yet, unknown.

Cellulose forms most of the fibre in a horse's diet. Nearly all farm foods contain cellulose. It is digested by the microbes in the hind gut, to produce volatile fatty acids which are used by the body cells as a source of energy. As plants age so the cellulose is converted to the completely indigestible material known as lignin. This is the skeleton of plants. So the later grass is left before cutting for hay the less digestible it is and the lower its nutritional value.

The horse relies on us to provide his mineral needs, through a well balanced ration and carefully managed pasture. To work correctly almost every body process requires various minerals. Minerals tend to interact with one another and cannot therefore be considered in isolation. A single dietary excess or deficiency can affect other available minerals and in turn many body functions in the horse.

These body functions also rely on the availability of one or more vitamins which in turn may interact with certain minerals. Vitamin activity deteriorates with time and is eventually destroyed by exposure to light, heat, moulds and oxidising agents. A vitamin/mineral supplement should be included in all home mixed straight rations but should not be added to bought-in balanced rations, otherwise in the latter case dangerous excesses could be created.

Salt is the main exception to this rule. It is recommended to provide lumps of rock-salt in all paddocks, and mineral-salt licks in all boxes.

It can further be advanced with certainty that the bulk of our home-produced fodder is definitely deficient in protein content. In my opinion, that may well be another important factor contributing to the low fertility of so many of our stud animals. If we consider the very poor, worn out, matty and horse-sick grazing on so many of our stud farms, we can only be reinforced in that opinion; young grass of the best quality, in the growing stage, may produce anything up to 15% digestible protein, compared to 60 or 70% of starchy matter. Whilst on the type of poor grazing referred to, the protein content may drop to 3% or to even less.

Accepting the fact, which is well established and undeniable, that most of our home-grown fodder is too low in protein content, we can only try and remedy the position by a continuous effort to produce, and to feed, as much protein-rich food as is possible under our climatic conditions.

To produce that effect, good first-class grazing is a mainstay. It is at its best in the spring, before it reaches the flowering stage and runs to seed. After that time its lignin content will increase considerably at the expense of digestible carbohydrates, and the protein content will be much lowered.

We can make up for the falling-off in feeding value of grass by providing other high-protein foods.

The first stand-by is lucerne: we may obtain as many as five consecutive cuts in a season, if used for grass feed only, or one cut for hay and two to three for aftermath, or two cuts for hay. Lucerne, known as alfalfa in America, is exceedingly rich in protein, and moreover its lime content is very high. It is the most valuable food we can grow, under suitable conditions, and no stud farm where it can be grown ought to be without it.

The next standby is maize, for green food. On fertile land, in the South of England, it produces a phenomenal amount of luscious and rich green food, available from the beginning of August till the end of September, just when the grazing is at its poorest. It is a rich, wholesome, succulent and very tasty food; the horses love it, and they thrive on it; it is easy and very clean to feed.

For the winter season, we must rely mainly on first-class hay. To be first class, it must be made of the right herbage, it must be cut at the right time, and it must be well-got. In my experience, I have never been able to buy really first-class hay. Bought hay is usually cut too late, when it contains too much indigestible fibre and has lost much of its protein; frequently it has been over exposed to the sun, is bleached, and has lost much valuable carotene. I have found it essential to make my own hay, and to make it in abundance.

There is no hay quite so valuable as very best quality lucerne hay. Such hay may contain as much as 19% crude protein, and 11 MJ/kg. digestible energy. In addition, this hay contains rather more than 1.75% of calcium. Sainfoin hay is nearly as good.

Seeds hay, made from a mixture of red clover and rye grasses, will contain about 12% of protein and 0.5% of calcium.

Clover hay, made from pure clovers, contains about 15% of protein and 1.6% of calcium.

Very good meadow hay, so-called soft hay, may contain as much as 13.5% crude protein and 9.7 MJ/kg. D.E.

Medium meadow hay, 8% protein and 8.5 MJ/kg. D.E.

Poor meadow hay, 4% protein and 6 MJ/kg. digestible energy, wherefrom it will be seen that poor hay has but a low feeding value.

Well-got hay is of a blue-green colour and possesses a sweet aroma.

Hay without aroma is a worthless feed; musty, fusty hay is injurious to health, and must on no account be fed.

Oats are the best energizing food for horses, so much better and more suitable than any other kind of grain that it is not worth considering any substitute. It is the proportion of protein, starch, fat, fibre and lime that makes oats so exceptionally wholesome and digestible to horses. Oats contain up to 11% crude protein, 12 MJ/kg. D.E., 4.5% oil, 10% fibre, and 3% minerals, including lime. The energizing elements, starch, fat, fibre and minerals form a very high percentage, but it will be noted that the protein/energy ratio is low and the protein is of poor quality in terms of essential amino acids, particularly lysine and meionine; it is necessary, therefore, particularly in the case of stud animals, to make good the protein deficiency of oats by feeding a high quality protein supplement such as extracted soyabean meal.

The woody fibre contained in the husk of oats adds greatly to the digestibility of this corn in the case of the horse. The horse needs a certain amount of fibre; he will get a considerable amount of fibre in his hay, since the percentage of fibre in hay varies from around 25 to nearly 35%. Even then, some horses crave for more fibre, and will eat their straw bed. Done in moderation, there is no harm in that; it is usually a sign of a healthy appetite.

The digestibility of oats is improved by the practice of bruising or rolling the grain, done only lightly, and just enough to crack the husk. This causes the horse to eat more slowly, and to masticate the grain more thoroughly; in so doing, he impregnates his food more thoroughly with saliva, which makes it more readily digestible as it enters the stomach. If oats are fed whole, a certain proportion of them will pass through the horse entirely undigested, as may be verified from his droppings. The addition of a handful of chaff to each oat feed causes the horse to chew still more carefully; chaff, or chop, may be cut from hay or from straw, or from a mixture of both.

Linseed should be available, and used in moderate quantities, on every stud. It is a food extraordinarily rich in fat or oil, of which it contains as much as 37%, but the protein is low quality and must not be relied upon to make good any deficiency in the ration. On account of its high oil content, it is most soothing to the digestive system, and is of great use in the case of ailing animals. It is a fattening food, a good conditioner, and it has a marked effect on the gloss of a horse's coat.

Linseed must not be fed except boiled, having been well soaked first. Soak the linseed overnight, well covered in water, allowing 0.5 to 0.7 kg. of seed for each horse. The next day, boil actively for 10 minutes and then simmer slowly for 2 or 3 hours, until the grain is thoroughly soft; linseed burns very easily, so stir frequently and use a moderate fire only. Now add it to the ration, stir and mix thoroughly, cover up, and let stand for about half an hour. The thicker this linseed mash is, the better the horses will eat it.

Alternatively, linseed jelly may be made, by stirring the soaked seed into boiling water at the rate of 4.5 kg. of linseed to 20 gallons of water. Then add a little cold water, bring to the boil again, and let boil for about 20 minutes. This jelly may now be mixed to oats, or to a mixture of food.

I believe it to be a good thing, in winter, to feed boiled, or preferably steamed, oats to all horses, and particularly to brood mares and young stock. During the short and cold winter days, when there is no grass about worth eating, horses turned out take very little exercise; they stand about most of the time. They are more inclined to be constipated than during the grazing season. I have found steamed oats to have a beneficial effect in this connection; if well done, they have a lovely aroma to them, they are a welcome change for the animals and, once they are used to them, are much appreciated. We may add some salt to the water used for steaming. Whereas boiling has the disadvantage of making the oats absorb too much water, which is not desirable, that trouble is overcome entirely by the use of a steam-boiler. With this we can steam anything up to 15 gallons of oats with as little as 3 to 4 gallons of water. Cooking improves the digestibility of whole grain by 3%. It ruptures the starch molecules as well as splitting the indigestible thick outer coat of the grain. It is a useful method of increasing the digestible energy of a ration, particularly in the case of old horses. However, cooking speeds up the destruction of some vitamins by oxidation. It destroys any mould spores and slightly increases the length of time it takes for the 'cooked' protein to be absorbed, which can be an advantage.

On a stud of some size, the bulk of food to be prepared is quite considerable, and when a mixture is required it is of importance to mix as thoroughly as possible. It cannot be done properly by mixing a bucket at a time, apart from the fact that to do so is a considerable waste of labour. It is essential to have a large mixing vessel. Even then there is still a lot of time and labour involved if the feed has to be carried bucket by bucket to each box. The best labour-saving arrangement is to have a feed-barrow. These can be purchased mounted on rubber tyres, with a content of from 50 to 80 gallons, which is sufficient for a considerable number of animals. Boiled food or mashes can be mixed directly in such a feed-barrow, and when ready, wheeled along the boxes. It is obvious that the possession of such an implement is a great saver of time and labour, apart from the fact that it ensures a thorough mixture. It will be understood that the feed-barrow must be cleaned out carefully each time after it has been used for boiled food or mashes. Any remains of damped food turn sour very quickly, and would spoil the next feed.

I am a great believer in having some natural fresh and succulent food to give to horses as part of their winter diet. For that purpose we may use carrots, turnips, mangolds or swedes. It is true that these roots contain nearly 90 per cent. water, only just over 1 per cent. protein, 1.76 MJ/kg. D.E., and practically no fat, fibre or minerals. None the less, they are to the horse what fruit is to us, and they quite obviously do him a lot of good. Besides, horses are very fond of them and eat them with real relish. That in itself is a good enough reason since it helps in keeping them happy and contented, and contentment is half the battle in maintaining condition and health. Carrots are usually fed sliced lengthwise; I feed turnips, mangolds or swedes whole.

I am not a believer in dosing horses too freely with medicines or with tonics of different kinds, which are usually expensive and frequently of doubtful value. However, the value of codliver oil as a tonic for cases of debility is so well established as to need no further argument. It has a beneficial influence on the formation of healthy bone, as a preventative for rickets, and also for pulmonary affections. The addition of one or two tablespoonfuls every day to the food of weaned foals, for instance, or to that of other horses a little low in condition, has much to recommend it. The only difficulty is that codliver oil possesses both a repulsive smell and a nasty taste, so that horses may easily be diffident about this admixture to their food. In that case malt and codliver oil may be the answer. Always buy oil in small quantities and store it in a cool, dark place. Rancidity is known to destroy vitamin E.

The one thing which is liable more than any other to pull a horse down in condition, is worm infestation. In my opinion, it is quite essential to have all inmates of a stud examined for worms at least once every spring, and to treat all animals, once every 4 to 6 weeks, throughout the year.

The horse is a particularly clean, and even dainty, feeder. Any kind of dirt in food, sourness, or untoward smell, puts him off his feed. Mangers must be kept scrupulously clean, and no traces of a previous feed allowed to remain; such remainders will ferment, and will sour, contaminate and spoil any subsequent meal. It is quite certain that it does make a lot of difference to present a horse's food in a pleasant and appetizing manner. Horses are just like ourselves in that respect. To be a good feeder is one of the highest praises that a stud-groom can earn for himself. It implies a good deal more than the ability to prepare a good feed. It implies an eye for condition, the ability to see from day to day which horses do well and which do not; horses vary a great deal amongst themselves, and that which suits one animal best may not benefit another anything like as well; horses, like human beings, vary from day to day, and it requires talent, observation and, above all, a love of animals and a real interest in them to do each individual justice.

An owner who possesses genuine interest in his animals cannot do better than be present frequently at evening stables. If everything is as it should be, he should find the inmates of his boxes intent and eager, and full of conversation, in anticipation of the arrival of their dinner. A little later he will find all noses well buried in their respective mangers, and dinner attended to with obvious enjoyment. It will give him an opportunity, without disturbing the diners, to have a good look at one or two of them, and to study their condition and well-being. In so doing, he will be able to listen to his stud-groom's comments, which are always interesting and which are bound to be given with pleasure. An intelligent interest in his work is one of the best incentives that can be given to any good man. The least thing that an owner should do is to see that his stud-groom need never be without any of the essential requisites of his trade, that is food of prime quality and of the necessary variety, medicines, tools and tack-room essentials.

The horse has a small stomach, which means that he must not be given a large feed at any one time; he should feed little and often. When at grass, feeding in the natural way, he sees to that very well himself, and will be feeding off and on most of the day. Kept in, the minimum that we should do, in order to approach his natural requirements, is to feed three times a day, seeing that he has some hay to play with during in-between times and at night.

But whilst the horse has a small stomach, he has an intestine of large capacity and requires a good deal of bulk, to be taken over a period, and gradually. This requirement is fulfilled in the main by grass and hay, and the horse cannot do without an adequate quantity of the one or the other.

Whilst I have tried in preceding chapters, dealing with different classes of stud animals, to give some idea regarding the quantities of various classes of food to be fed to them, this can be taken as a rough and ready guide only. The description of a good feeder, in this chapter, ought to have made it clear that no hard and fast rules can be made, since the best answer to each individual's requirements may vary from day to day. But one rule can be made, to the effect that no horse should be fed, at any one time, a larger quantity of concentrated food, such as oats, than he will clear up readily. This rule does not apply to bulky food such as hay, which is there for the very purpose of being consumed slowly, at leisure, and at the horse's will.

And, when all is said and done, remember, it is the master's eye which makes the horse fat!

GESTATION TABLE

Date of Service	Birth	Date of Service	Birth	Date of Service	Birth	Date of Service	Birth
Jan. 1	Dec. 6	Apr. 1	Mar. 6	July 1	June 5	Oct. 1	Sept. 5
2	7	2	7	2	6	2	6
4	9	4	9	4	8	4	8
6	11	6	11	6	10	6	10
8	13	8	13	8	12	8	12
10	15	10	15	10	14	10	14
12	17	12	17	12	16	12	16
14	19	14	19	14	18	14	18
16	21	16	21	16	20	16	20
18	23	18	23	18	22	18	22
20	25	20	25	20	24	20	24
22	27	22	27	22	26	22	26
24	29	24	29	24	28	24	28
26	31	26	31	26	30	26	30
28	Jan. 2	28	Apr. 2	28	July 2	28	Oct. 2
30	4	30	4	30	4	30	4
Feb. 1	Jan. 6	May 1	Apr. 5	Aug. 1	July 6	Nov. 1	Oct. 6
2	7	2	6	2	7	2	7
4	9	4	8	4	9	4	9
6	11	6	10	6	11	6	11
8	13	8	12	8	13	8	13
10	15	10	14	10	15	10	15
12	17	12	16	12	17	12	17
14	19	14	18	14	19	14	19
16	21	16	20	16	21	16	21
18	23	18	22	18	23	18	23
20	25	20	24	20	25	20	25
22	27	22	26	22	27	22	27
24	29	24	28	24	29	24	29
26	31	26	30	26	31	26	31
28	Feb. 2	28	May 2	28	Aug. 2	28	Nov. 2
		30	4	30	4	30	4
Mar. 1	Feb. 3	June 1	May 6	Sept. 1	Aug. 6	Dec. 1	Nov. 5
2	4	2	7	2	7	2	6
4	6	4	9	4	9	4	8
6	8	6	11	6	11	6	10
8	10	8	13	8	13	8	12
10	12	10	15	10	15	10	14
12	14	12	17	12	17	12	16
14	16	14	19	14	19	14	18
16	18	16	21	16	21	16	20
18	20	18	23	18	23	18	22
20	22	20	25	20	25	20	24
22	24	22	27	22	27	22	26
24	26	24	29	24	29	24	28
26	28	26	31	26	31	26	30
28	Mar. 2	28	June 2	28	Sept. 2	28	Dec. 2
30	4	30	4	30	4	30	4

Index